KEW BULLETIN 56: 257 – 360 (2001)

Cyperaceae of economic, ethnobotanical and horticultural importance: a checklist

DAVID A. SIMPSON[1] & CECILIA A. INGLIS[2]

Summary. A checklist of *Cyperaceae* recorded as having economic, ethnobotanical and horticultural importance is presented. Data have been obtained from the literature and herbarium material. Forty-five genera and 502 species/infraspecific taxa are included.

CONTENTS

INTRODUCTION

Cyperaceae are the third largest family in the Monocotyledons and comprise c. 104 genera and c. 5000 species (Goetghebeur 1998). They have a cosmopolitan distribution, with a concentration of genera in the tropics. The largest genus is *Carex*, with c. 2000 species, followed by *Cyperus* with c. 600 species; *Cyperaceae* are of economic significance worldwide. Forty-five genera and 502 species/infraspecific taxa are identified as such here. Their importance is often at a regional or local level and the family plays a vital part in many local economies. It is probably due to their localised use that they have generally been overlooked as plants of economic importance. Moreover, relevant data are scattered through the literature or on herbarium specimens. The aim of this work is twofold: a) to bring together the disparate data in the form of a checklist and b) to increase awareness of the economic value of this taxonomically difficult but highly interesting family. The checklist was compiled from a survey of relevant literature and herbarium specimens in Kew and elsewhere. They were entered on to an ALICE format database using the first and second level descriptors for economic use delimited by Cook (1995). Each entry in the checklist comprises the accepted name, basionym and important synonyms where these are considered necessary (for instance if a species is often recognised in another genus),

Accepted for publication February 2001.
[1] Herbarium, Royal Botanic Gardens, Kew, Richmond, Surrey, TW9 3AB, U.K.
[2] 22 Hollingbourne Avenue, Bexleyheath, Kent, DA7 5EU, U.K.

general distribution, habitat and then economic/ethnobotanical importance. We have also included 'Weeds' and 'Active Compounds Present' as separate first level descriptors. In the case of the former, many *Cyperaceae* are considered to be serious weeds and their presence as such is of particular economic importance in some regions. Interestingly, many species which are considered to be weeds also have uses. Further brief information is provided under the second level descriptor, if such information is available. This includes the country from which the data were recorded, if known. Reference citations beginning with 'H' followed by a number represent a herbarium specimen. Infraspecific taxa have been recognised when there is specific reference to one in the data source. Otherwise, data have been assembled under species names. Following the main Checklist, a further list arranges taxa under first and second level descriptor headings. This will allow rapid determination of the names and number of taxa assigned to particular descriptors. As far as possible we have checked and verified the names used; we have not included records where the naming is uncertain or incorrect. We have taken a global view in this Checklist and have endeavoured to include as many taxa as possible. However, our search for records has been limited by time and no doubt we have overlooked taxa, especially from the New World where both the literature and herbarium material are less well known to us. The first author would welcome any new data or confirmation of the data given below.

THE CHECKLIST

ACTINOSCIRPUS (*Ohwi*) *R. W. Haines & Lye*

Actinoscirpus grossus (*L. f.*) *Goetgh. & D. A. Simpson*, Kew Bull. 46: 171 (1991).
Scirpus grossus L.f., Suppl.: 104 (1781).
India, SE Asia; swamps and cultivated areas, particularly rice fields.

MATERIALS: *Fibres*: mats, bags and baskets. Culms cut, split, dried, flattened and sun-bleached; may be dyed, Thailand, Malaysia (Kern 1974; Simpson 1992a).

MEDICINES: *Unspecified medicinal disorders*: affects both heat production and heat regulation in the body, India (Cauis & Banby 1935); *Digestive system disorders*: vomiting, diarrhoea, India (Cauis & Banby 1935; Heywood 1993); *Ill-defined symptoms*: liver tonic, India (Cauis & Banby 1935); *Infections/infestations*: useful against infection and poisons, including gonorrhoea and leprosy, India (Cauis & Banby 1935); *Nutritional disorders*: appetite stimulant; improves sense of taste, India (Cauis & Banby 1935); *Sensory system*: eye problems, India (Cauis & Banby 1935); *Skin and subcutaneous cellular tissue disorders*: useful for treatment of burning sensations, India (Cauis & Banby 1935).

ENVIRONMENTAL USES: *Soil improvers*: ploughed in as green manure, SE Asia (Burkill 1935).

WEEDS: *Rice fields*: (Simpson & Koyama 1998).

var. **kysoor** (*Roxb.*) *Noltie*, Edinburgh J. Bot. 51 (2): 173 (1994).
Scirpus kysoor Roxb., Hort. Bengal.: 6 (1814).
Pakistan, India, Malay Peninsula; swamps, cultivated areas.
FOOD: *Tubers*: (Burkill 1935).

MATERIALS: *Fibres*: culms used for string and matting, Malaysia (Burkill 1935).

MEDICINES: *Unspecified medicinal disorders*: tubers astringent, Malaysia (Burkill 1935); *Digestive system disorders*: used to relieve diarrhoea and vomiting (Millar & Morris 1988).

AFROTRILEPIS (*Gilly*) *J. Raynal*

Afrotrilepis pilosa (*Boeck.*) *J. Raynal*, Adansonia sér. 2, 3: 258 (1963).
Trilepis pilosa Boeck., Linnaea 39: 10 (1875).
West tropical Africa from Senegal to Gabon; granite outcrops.

MATERIALS: *Fibres*: thatching, Gabon and Sierra Leone (Burkill 1985).

SOCIAL USES: '*Religious*' *uses*: tuft placed on top of huts said to ward off lightning, Ivory Coast (Burkill 1985).

ASCOLEPIS *Nees ex Steud.*

Ascolepis capensis (*Kunth*) *Ridl.*, Trans. Linn. Soc. London, Bot. 2: 164 (1884).
Platylepis capensis Kunth, Enum. Pl. 2: 269 (1837).
Widespread in tropical Africa from Mali to Ethiopia and to S Africa; seasonally or permanently wet grasslands and swamp margins.

ANIMAL FOOD: *Unspecified part*: grazing for cattle, Ethiopia (Burkill 1985).

BOLBOSCHOENUS (*Asch.*) *Palla*

Bolboschoenus caldwellii (*V. J. Cook*) *Soják*, Čas. Nár. Mus., Odd. Přír. 141 (1 – 2): 62 (1972).
Scirpus caldwellii V. J. Cook, Trans. & Proc. Roy. Soc. New Zealand 76: 568 (1947).
SE Australia and New Zealand; coastal saline and sandy areas, drains and waterways.

ANIMAL FOOD: *Leaves/culms/aerial parts*: Australia (Anon. 1994 – 2000).

WEEDS: *Aquatic biotopes*: (Moore & Edgar 1980); *Irrigation ditches*: (Moore & Edgar 1980).

Bolboschoenus fluviatilis (*Torr.*) *Soják*, Čas. Nár. Mus., Odd. Přír. 141 (1 – 2): 62 (1972).
Scirpus maritimus L. var. *fluviatilis* Torr., Ann. Lyceum Nat. Hist. New York 3: 24 (1836).
N America, E Asia, Australia, New Zealand; river and stream margins.

FOOD: *Rhizomes*: inner portions used to provide an occasional food source for Maori, New Zealand (Johnson 1989); possibly planted as food source by Maori, New Zealand (H175).

Bolboschoenus maritimus (*L..*) *Palla* in Koch, Syn. Deut. Schweiz. Fl. ed. 3: 2532 (1904).
Scirpus maritimus L., Sp. Pl. 1: 51 (1753).
± Cosmopolitan, except for Arctic regions; seasonally or permanently wet areas on saline soils often near coasts.

FOOD: *Rhizomes*: cooked vegetable (Anon. 1994 – 2000).

MATERIALS: *Fibres*: culms used for thatching, Ethiopia (Burkill 1985); *Other materials/chemicals*: culms used for locust-bait, Somalia (Burkill 1985).

MEDICINES: *Unspecified medicinal disorders*: astringent, China (Cauis & Banby 1935); *Genito-urinary system disorders*: diuretic, China (Cauis & Banby 1935).

ENVIRONMENTAL USES: *Ornamentals*: (Huxley 1992; Royal Horticultural Society 2000).

WEEDS: *Aquatic biotopes*: (Kühn 1982); *Rice fields*: (Kühn 1982).

ACTIVE COMPOUNDS PRESENT: *Alkaloids*: (Burkill 1985).

Bolboschoenus paludosus (*A. Nelson*) *Soó*, Acta Bot. Acad. Sci. Hung. 16 (3 – 4): 368 (1970, publ. 1971).
Scirpus paludosus A. Nelson, Bull. Torrey Bot. Club 26: 5 (1899).
N and C America, Argentina, Hawaii; saline areas, flooded soils.

FOOD: *Nutlets*: edible (Anon. 1994 – 2000).

Bolboschoenus yagara (*Ohwi*) *Y. C. Yang & M. Zhan*, Acta Biol. Plateau Sin. 7: 14 (1987, publ. 1988).
Scirpus yagara Ohwi, Mem. Coll. Sci. Kyoto Imp. Univ., Ser. B, Biol. 8, 1: 110 (1944).
E Asia; wet places.

MEDICINES: *Blood system disorders*: rhizomes used to reduce blood clotting, China (J. C. Shaw, C. S. Tong & L. Wong, pers. comm. 1995).

BULBOSTYLIS *Kunth*

Bulbostylis barbata (*Rottb.*) *C. B. Clarke* in Hook. f., Fl. Brit. India 6: 651 (1893).
Scirpus barbatus Rottb., Desc. Pl. Rar.: 27 (1772); Descr. Icon. Rar. Pl.: 52, t. 17, f. 4 (1773).
Old World tropics and subtropics, SE U.S.A.; open sandy areas and waste places, grasslands and rocky outcrops, often coastal.

ANIMAL FOOD: *Unspecified part*: grazed by sheep and goats, but poor value, India (Kulhari & Joshi 1992).

WEEDS: *Cultivation*: (Burkill 1985; Dangol 1992); *Rotation crops*: (Kühn 1982); soya bean fields, Thailand (Radanachaless & Maxwell 1994); *Perennial crops*: (Kühn 1982); *Waste places*: (Kühn 1982).

Bulbostylis capillaris (*L.*) *C. B. Clarke* in Hook. f., Fl. Brit. India 6: 652 (1893).
Scirpus capillaris L., Sp. Pl. 1: 49 (1753).
Tropical and subtropical America, Caribbean; open, often sandy areas, rocky outcrops, grasslands.

MEDICINES: *Genito-urinary system disorders*: blood purifier in female disorders, Paraguay (Altschul 1973); *Pain*: mouthwash for toothache, Paraguay (Altschul 1973).

Bulbostylis densa (*Wall.*) *Hand.-Mazz.*, Vegetationsbilder 20 (7): 16 (1930).
Scirpus densus Wall. in Roxb., Fl. Ind. 1: 231 (1820).
Widespread throughout the Old World tropics; open, grassy areas, rocky outcrops,

margins of water bodies.

WEEDS: *Rotation crops*: (Kühn 1982); *Perennial crops*: (Kühn 1982); *Grassland*: (Kühn 1982); *Waste places*: (Kühn 1982); *Aquatic biotopes*: (Kühn 1982).

Bulbostylis hispidula (*Vahl*) *R. W. Haines* in R. W. Haines & Lye, Sedges & Rushes E. Afr. App. 3: 1 (1983).
Scirpus hispidulus Vahl, Enum. Pl. 2: 276 (1806).
Fimbristylis hispidula (Vahl) Kunth, Enum. Pl. 2: 227 (1837).
Tropical Africa; dry, sandy places, shallow soil over rocks, disturbed places.

ANIMAL FOOD: *Unspecified part*: grazed by cattle when nothing else is available, Senegal (Burkill 1985).

MATERIALS: *Fibres*: sometimes used in matting (Burkill 1985).

MEDICINES: *Skin/subcutaneous cellular tissue disorders*: whole plant burnt with *Cordia africana* Lam. to fumigate fingers affected by handling cotton thread (Burkill 1985).

WEEDS: *Cultivation*: (Burkill 1985); *Perennial crops*: (Kühn 1982); *Grassland*: (Kühn 1982); *Aquatic biotopes*: (Kühn 1982).

Bulbostylis junciformis (*Kunth*) *C. B. Clarke ex S. Moore*, Trans. Linn. Soc. London, Bot. 4: 512 (1895).
Isolepis junciformis Kunth in Humb., Bonpl. & Kunth., Nov. Gen. Sp. 1: 222 (1816).
C America, W Indies, tropical S America; wet sandy soils or savannahs.

MEDICINES: *Unspecified medicinal disorders*: decoction of leaves and culm used by Tiriyó as external bath, Brazil (Milliken 1997).

Bulbostylis lanata *C. B. Clarke*, Kew Bull. Addit. Ser. 8: 26 (1908).
Isolepis lanata Humb., Bonpl. & Kunth, Nov. Gen. & Sp. 1: 220 (1815).
Tropical S America, especially Brazil; seasonally wet areas in savannah

MEDICINES: *Unspecified medical disorders*: decoction of whole used by Tiriyó as external bath for fevers with headaches, Brazil (Milliken 1997).

Bulbostylis metralis *Cherm.*, Arch. Bot. Mém. 7: 34 (1931).
Tropical Africa; seasonally wet habitats.

MATERIALS: *Other materials/chemicals*: whole plant used to make brushes, Ghana (Abbiw 1990).

Bulbostylis pilosa (*Willd.*) *Cherm.*, Bull. Soc. Bot. France 81: 266 (1934).
Schoenus pilosus Willd., Phytographia 1: 3 (1794).
Tropical Africa; open grasslands.

MATERIALS: *Other materials/chemicals*: culms and whole plant used to make brooms, Ghana (Abbiw 1990; Burkill 1985).

Bulbostylis puberula (*Poir.*) *C. B. Clarke* in Hook. f., Fl. Brit. India 6: 652 (1893).
Scirpus puberulus Poir. in Lam., Encycl. 6: 767 (1805).
Tropical Africa and Asia; dry sandy areas.

MEDICINES: *Genito-urinary system disorders*: used as a diuretic, has been cultivated in Singapore by Chinese and exported to China (Kern 1974).

CAREX *L.*

Carex acuta *L.*, Sp. Pl. 1: 975 (1753).
Europe, N and W Asia, N Africa and N America; ponds, riversides and marshy places.
    ENVIRONMENTAL USES: *Ornamentals*: ground cover (Grounds 1989; Huxley 1992).

Carex acutiformis *Ehrh.*, Beitr. Naturk. 4: 43 (1789).
Europe, N Africa, temperate Asia, N America; swampy areas and wet grasslands.
    ENVIRONMENTAL USES: *Ornamentals*: ground cover (Huxley 1992).

Carex alba *Scop.*, Fl. Carniol. ed. 2, 2: 216 (1772).
Europe, N Asia; grasslands, boggy areas, woodlands.
    ENVIRONMENTAL USES: *Ornamentals*: ground cover (Huxley 1992).

Carex albonigra *Mack.* in Rydb., Fl. Rocky Mts.: 137 (1917).
Canada and U.S.A.; high mountain slopes and summits.
    ANIMAL FOOD: *Unspecified part*: moderately grazed by sheep, U.S.A. (Hermann 1970).

Carex alboviridis *C. B. Clarke*, J. Linn. Soc., Bot. 29: 62 (1891).
Madagascar; dry sandy areas.
    MEDICINES: *Unspecified medicinal disorders*: Madagascar (Cauis & Banby 1935).

Carex albula *Allan*, Trans. & Proc. Roy. Soc. New Zealand 76: 589 (1947).
New Zealand; very dry grasslands and cultivated areas
    ENVIRONMENTAL USES: *Ornamentals*: ground cover (Grounds 1989).
    WEEDS: *Rotation crops*: (Healy & Edgar 1980); *Grassland*: (Healy & Edgar 1980).

Carex amplifolia *Boott* in Hook., Fl. Bor.-Amer. 2: 228 (1839).
Canada and U.S.A.; wet soil along streams and in meadows.
    ANIMAL FOOD: *Unspecified part*: grazed by cattle and horses, U.S.A. (Hermann 1970).

Carex appressa *R. Br.*, Prodr.: 242 (1810).
Malesia, New Caledonia, Australia and New Zealand; marshes, lake shores and meadows.
    ANIMAL FOOD: *Leaves/culms/aerial parts*: leaves (Anon. 1994 – 2000).
    MATERIALS: *Fibres*: leaves made into fibre baskets (Anon. 1994 – 2000).
    ENVIRONMENTAL USES: *Erosion control*: live plant (Anon. 1994 – 2000).

var. **secta** (*Boott*) *Kük.*, in Engl., Pflanzenr. 4 (20), 38 Heft: 179 (1909).
Carex secta Boott in Hook. f., Fl. Nov.-Zel. 1: 281 (1853).
New Zealand; swamps.
    ENVIRONMENTAL USES: *Ornamentals*: ground cover (Huxley 1992).

Carex aquatilis *Wahl.*, Kongl. Vetensk. Acad. Nya Handl. 24: 165 (1803).
N America, Greenland, Eurasia; swampy areas, often in shallow water.

ANIMAL FOOD: *Unspecified part*: medium to high forage value, N America (Hermann 1970); in arctic regions reported to be outstanding as forage, equalling clover in nutritional value and exceeding it in protein content (Hermann 1970); a common component of meadow hay in many areas, N America (Hermann 1970).

Carex arenaria *L.*, Sp. Pl. 1: 973 (1753).
N and E Europe, Siberia, N America; coastal sandy areas.
    MEDICINES: *Unspecified medicinal disorders*: Europe (Cauis & Banby 1935).
    ENVIRONMENTAL USES: *Ornamentals*: ground cover (Huxley 1992).

Carex athrostachya *Olney*, Proc. Amer. Acad. Arts 7: 393 (1868).
Mountain valleys to the spruce-fir zone; wet meadows and thickets.
    ANIMAL FOOD: *Unspecified part*: fair to good forage value for horses and cattle, U.S.A. (Hermann 1970).

Carex atrata *L.*, Sp. Pl. 2: 976 (1753).
Europe, N America; wet, rocky areas, lowland in Arctic regions, mountains elsewhere.
    ENVIRONMENTAL USES: *Ornamentals*: ground cover (Grounds 1989; Huxley 1992).

Carex baccans *Nees* in Wight, Contrib. Bot. India: 122 (1834).
India to Taiwan, Malesia; open grassy slopes, scrub, forest margins and clearings.
    ENVIRONMENTAL USES: *Ornamentals*: (Huxley 1992).

Carex baldensis *L.*, Cent. Pl. 2: 32 (1756).
Europe (SE Alps); dry, rocky places.
    ENVIRONMENTAL USES: *Ornamentals*: (Huxley 1992).

Carex bella *L. H. Bailey*, Bot. Gaz. (Crawfordsville) 17: 152 (1892).
U.S.A.; along streams and in moist open woods and open parks.
    ANIMAL FOOD: *Unspecified part*: highly palatable to all livestock and, where locally abundant, an important forage sedge, U.S.A. (Hermann 1970).

Carex berggrenii *Petrie*, Trans. & Proc. New Zealand Inst. 18: 297 (1886).
New Zealand; boggy ground, river flats or lake shores.
    ENVIRONMENTAL USES: *Ornamentals*: (Grounds 1989; Huxley 1992).

Carex bigelowii *Torr.*, Ann. Lyceum Nat. Hist. New York 1: 67 (1824).
N America, Greenland, arctic and alpine Eurasia; mountain meadows, gravelly slopes, open rocky shores.
    ANIMAL FOOD: *Unspecified part*: fair to excellent palatability for livestock, U.S.A. (Hermann 1970).

Carex breviculmis *R. Br.*, Prodr.: 242 (1810).
SE Asia, Australia, New Zealand; grasslands, open rocky places.
    WEEDS: *Grassland*: New Zealand (Healy & Edgar 1980); *Gardens*: New Zealand (Healy & Edgar 1980).

Carex brevior *(Dewey) Mack.*, Amer. Midl. Naturalist 4: 235 (1915).
Canada and U.S.A.; open plains and moist place in mountains.
    ANIMAL FOOD: *Unspecified part*: medium forage value for all livestock, U.S.A. (Hermann 1970).

Carex brevipes *W. Boott* in S. Watson, Bot. California 2: 246 (1880).
U.S.A.; dry, open woods, clearings and slopes.
    ANIMAL FOOD: *Unspecified part*: good for cattle and horses, fair for sheep, U.S.A. (Hermann 1970).

Carex brizoides *L.*, Cent. Pl. 1: 31 (1755).
C Europe; damp, shady places.
    MATERIALS: *Fibres*: woven into mats, Switzerland (H284 — Fig. 1D); *Other materials/ chemicals*: culms and leaves, packing material (Heywood 1993).
    WEEDS: *Perennial crops*: (Kühn 1982); *Grassland*: (Kühn 1982).

Carex brunnea *Thunb.*, Fl. Jap.: 38 (1784).
Madagascar to India, Japan, New Caledonia, Australia; marshy areas, margins of water bodies, alpine grassland.
    ENVIRONMENTAL USES: *Ornamentals*: (Grounds 1989; Huxley 1992).

Carex buchananii *Berggr.*, J. Bot. 18: 104 (1880).
New Zealand; coastal scrub, damp ground near streams and short-tussock grassland.
    ENVIRONMENTAL USES: *Ornamentals*: (Grounds 1989; Huxley 1992).
    WEEDS: *Grassland*: (Healy & Edgar 1980; Parsons & Cuthbertson 1992).

Carex canescens *L.*, Sp. Pl. 1: 974 (1753).
N America, Eurasia, Australia; lake margins, swamps, bogs.
    ANIMAL FOOD: *Unspecified part*: high palatability to cattle, U.S.A. (Hermann 1970).

Carex caryophyllea *Latourr.*, Chlor. Lugd.: 27 (1785).
Europe; dry, grassy or stony places.
    ENVIRONMENTAL USES: *Ornamentals*: (Huxley 1992).

Carex cernua *Boott*, Illustr. *Carex* 4: 171, t. 578 (1867).
India, China, Japan; margins of water bodies.
    VERTEBRATE POISONS: *Mammals*: cattle, causes lack of appetite, loss of milk and nervous symptoms, India (Cauis & Banby 1935).

Carex chlorosaccus *C. B. Clarke*, J. Linn. Soc., Bot. 34: 298 (1899).
Tropical Africa; moist forest sides of roads and paths.
    ANIMAL FOOD: *Unspecified part*: grazing, domestic livestock & buffalo, Kenya (Burkill 1985).

Carex comans *Berggr.*, Minneskr. Fisiogr. Sällsk. Lund, Art. 8: 28, t. 7, f. 15 – 19 (1878).
New Zealand; grasslands, pastures and river flats.

ENVIRONMENTAL USES: *Ornamentals*: ornamental in pebble gardens, sold as 'frosted curls', Australia, New Zealand (Grounds 1989; Huxley 1992; Moore & Edgar 1980; Parsons & Cuthbertson 1992).

WEEDS: *Rotation crops*: (Healy & Edgar 1980); *Grassland*: (Healy & Edgar 1980); *Gardens*: (Healy & Edgar 1980); *Pasture*: (Healy & Edgar 1980; Parsons & Cuthbertson 1992).

Carex conica *Boott ex A. Gray*, Narr. Exped. Perry 2: 325 (1857).
Japan, S Korea; wet places.
ENVIRONMENTAL USES: *Ornamentals*: (Grounds 1989; Huxley 1992).

Carex coriacea *Hamlin*, Trans. Roy. Soc. New Zealand 82: 63 (1954).
New Zealand; seepages, damp grassland, swampy river flats.
ANIMAL FOOD: *Leaves/culms/aerial parts*: young shoots grazed by cattle (Moore & Edgar 1970).
WEEDS: *Grassland*: one of the most troublesome and weedy *Carex* species in New Zealand (Healy & Edgar 1980).

FIG. 1. **A** *Machaerina rubiginosa* bag, Australia (EBMC 34381); **B** *Schoenus melanostachys* basket, Australia (EBMC 34393); **C** *Carex paniculata* hassock, U.K. (EBMC 34198); **D** *Carex brizoides* mat, Switzerland (EBMC 34193); **E** *Eleocharis* sp. saddle pad, Brazil (EBMC 73338); **F** *Cyperus alternifolius* dish, Ethiopia (EBMC 73595); **G** *Cyperus alternifolius* bread basket (EBMC 73609). **H** *Schoenoplectus californicus* mat, Ecuador (EBMC 40495). All from the Economic Botany Museum collections, Kew; catalogue number in brackets.

Carex curvula *All.*, Fl. Pedem. 2: 264, t. 92, f. 3 (1785).
C and S Europe; grassy and rocky areas.
ENVIRONMENTAL USES: *Ornamentals*: (Huxley 1992).

Carex demissa *Hornem.*, Fors. Oecon. Plantel 3, 1: 939 (1821).
Europe; introduced into New Zealand; damp places, usually on acid soils.
ANIMAL FOOD: *Unspecified part*: grazing for cattle and sheep, New Zealand (Healy & Edgar 1980).

Carex diandra *Schrank.*, Act. Acad. Elect. Mogunt. Sci. Util. Erfuti 3: 49 (1782).
Europe, N Asia, N America; woodland and dry rocky areas.
ENVIRONMENTAL USES: *Ornamentals*: (Huxley 1992).

Carex digitata *L.*, Sp. Pl. 2: 975 (1753).
Europe; woodland and dry rocky areas.
ENVIRONMENTAL USES: *Ornamentals*: (Huxley 1992).

Carex dipsacea *Berggr.*, Minneskr. Fisiogr. Sällsk. Lund, Art. 8: 28, t. 7, f. 8 – 14 (1878).
New Zealand; damp and swampy areas, tussock grassland.
ENVIRONMENTAL USES: *Ornamentals*: (Grounds 1989; Huxley 1992).

Carex dispalata *Boott ex A. Gray*, Narr. Exped. Perry 2: 325 (1857).
China, Japan; swampy areas.
MATERIALS: *Fibres*: cultivated, leaves used to make hats, Japan (Heywood 1993).

Carex dissita *Sol. ex Boott* in Hook. f., Fl. Nov.-Zel. 1: 284 (1853).
New Zealand; forest, scrub or swampy areas.
ENVIRONMENTAL USES: *Ornamentals*: (Huxley 1992).

Carex disticha *Huds.*, Fl. Angl.: 347 (1762).
Europe except Mediterranean and far north; damp meadows and grassland.
MEDICINES: *Unspecified medicinal disorders*: Europe (Cauis & Banby 1935).

Carex divulsa *Stokes* in With., Bot. Arr. Brit. Pl. ed. 2, 2: 1035 (1787).
Europe except far north. Introduced to Australia and New Zealand; grassy areas, often shaded.
ANIMAL FOOD: *Unspecified part*: cattle, New Zealand (Healy & Edgar 1980).
WEEDS: *Grassland*: New Zealand (Healy & Edgar 1980).

subsp. **leersii** (*Kneuck.*) *W. Koch*, Mitt. Bad. Landesvereins Naturk. Naturschutz Freiburg 11: 259 (1923).
Carex muricata L. var. *leersii* Kneuck. in Seubert, Excurs.-Fl. Baden: 52 (1891).
Europe; dry places on alkaline soils.
ENVIRONMENTAL USES: *Ornamentals*: (Grounds 1989).

Carex douglasii *Boott.* in Hook., Fl. Bor.-Amer.. 2: 213 (1839).
N America; dry, often alkaline, open plains, foothills and lower mountains.
ANIMAL FOOD: *Unspecified part*: important forage plant withstanding heavy grazing, N America (Hermann 1970).

Carex ebenea *Rydb.*, Bull. Torrey Bot. Club 28: 266 (1901).
U.S.A.; mountain meadows and clearings.
ANIMAL FOOD: *Unspecified part*: medium forage value to cattle and horses, less to sheep, U.S.A. (Hermann 1970).

Carex egglestonii *Mack.*, Bull. Torrey Bot. Club 42: 614 (1915).
U.S.A.; dry open soil in mountains.
ANIMAL FOOD: *Unspecified part*: important forage plant, where abundant, for cattle, U.S.A. (Hermann 1970).

Carex elata *All.*, Fl. Pedem. 2: 272 (1785).
Carex hudsonii A. Benn. in Watson, Lond. Cat. Brit. Pl. 9: 41 (1895).
Europe; ditches and wet places by rivers and lakes.
ENVIRONMENTAL USES: *Ornamentals*: (Grounds 1989; Huxley 1992).

Carex eleocharis *L. H. Bailey*, Mem. Torrey Bot. Club 1: 6 (1889).
N America; dry plains and foothills.
ANIMAL FOOD: *Unspecified part*: low to fair forage value for cattle and horses, its resistance to grazing particularly valued in overgrazed areas, U.S.A. (Hermann 1970).

Carex elynoides *Holm*, Amer. J. Sci. 4, 9: 356 (1900).
U.S.A.; mountain summits and alpine slopes.
ANIMAL FOOD: *Unspecified part*: fair to excellent forage value for livestock, especially sheep, U.S.A. (Hermann 1970).

Carex eurycarpa *Holm*, Amer. J. Sci. 4, 20: 303 (1905).
U.S.A.; wet meadows and stream banks.
ANIMAL FOOD: *Unspecified part*: excellent forage for cattle and horses less palatable to sheep, U.S.A. (Hermann 1970).

Carex exserta *Mack.*, Bull. Torrey Bot. Club. 42: 620 (1915).
N America; damp places in montane regions.
ENVIRONMENTAL USES: *Revegetators*: restoration of vegetation cover using sod plugs, U.S.A. (Ratliff & Westfall 1992).

Carex festivella *Mack.*, Bull. Torrey Bot. Club 42: 609 (1915).
N America; meadows and open slopes mostly montane.
ANIMAL FOOD: *Unspecified part*: medium forage value to cattle and horses, U.S.A. (Hermann 1970).

Carex filicina *Nees* in Wight, Contrib. Bot. India: 123 (1834).
India and Sri Lanka to China and Taiwan, Malesia; forest openings, grassy areas, river banks, alpine shrub.
> FOOD: *Nutlets*: eaten raw, India (Pal 1992b).

Carex filifolia *Nutt.*, Gen. N. Amer. Pl. 2: 204 (1818).
N America; dry stony hillsides and pastures.
> ANIMAL FOOD: *Unspecified part*: cattle, U.S.A. (Adams *et al.* 1989); good to excellent forage value for horses and sheep (Hermann 1970).

Carex firma *Host*, Syn. Pl.: 509 (1797).
C Europe; stony grassland and rocks.
> ENVIRONMENTAL USES: *Ornamentals*: (Grounds 1989; Huxley 1992).

Carex flacca *Schreb.*, Spic. Fl. Lips., App.: 669 (1771).
Europe, N Africa, Siberia; introduced into N America, Caribbean and New Zealand; calcareous grassland, woods, fens and bogs.
> ENVIRONMENTAL USES: *Ornamentals*: (Grounds 1989; Huxley 1992).

Carex flagellifera *Colenso*, Trans. & Proc. New Zealand Inst. 16: 342 (1884).
Australia and New Zealand; damp areas.
> WEEDS: *Pasture*: (Healy & Edgar 1980; Parsons & Cuthbertson 1992).

Carex foenea *Willd.*, Enum. Pl.: 957 (1809).
C and northern N America; dry, open areas, sometimes montane.
> ANIMAL FOOD: *Unspecified part*: moderate forage value for cattle and horses when in sufficient quantities, N America. (Hermann 1970).

Carex gaudichaudiana *Kunth*, Enum. Pl. 2: 417 (1837).
New Guinea, Australia, New Zealand; boggy ground.
> ENVIRONMENTAL USES: *Ornamentals*: (Huxley 1992).

Carex geyeri *Boott*, Trans. Linn. Soc. London 20: 118 (1846).
Canada and U.S.A.; open woodland and dry slopes at middle and high elevations.
> ANIMAL FOOD: *Unspecified part*: readily grazed by cattle, U.S.A. (Hermann 1970).
> ENVIRONMENTAL USES: *Erosion control*: protects loose granitic soils, damage to which has caused serious watershed problems, U.S.A. (Hermann 1970).

Carex grayi *J. Carey* in A. Gray, Manual 1: 563 (1848).
Eastern N America; swamps, wet wooded areas.
> ENVIRONMENTAL USES: *Ornamentals*: (Huxley 1992).

Carex hachijoensis *Akiyama*, J. Jap. Bot. 13: 645 (1937).
Japan; not known.
> ENVIRONMENTAL USES: *Ornamentals*: (Grounds 1989; Huxley 1992).

Carex haydeniana *Olney* in S. Watson, Botany [fortieth parallel]: 366 (1871).
N America; alpine and subalpine slopes, and clearings.
 ANIMAL FOOD: *Unspecified part*: often rated as having high forage value due to the abundant fruit, U.S.A. (Hermann 1970).

Carex helophila *Mack.*, Torreya 13: 15 (1913).
Canada and U.S.A.; prairies, plains and hills.
 ANIMAL FOOD: *Unspecified part*: 50 – 80% palatable to all livestock, important at beginning of grazing season and after rains, U.S.A. (Hermann 1970).

Carex hepburnei *Boott* in Hook., Fl. Bor.-Amer. 2: 209 (1839).
Western N America; dry alpine summits and slopes, rarely streamside meadows.
 ANIMAL FOOD: *Unspecified part*: low forage value but grazed by livestock, N America (Hermann 1970).

Carex heteroneura *S. Watson*, Bot. California 2: 40 (1880).

var. **epapillosa** (*Mack.*) F. J. Herm., Rhodora 70: 421 (1968).
Carex epapillosa Mack. in Rydb., Fl. Rocky Mts.: 138 (1917).
Canada and U.S.A.; mountain meadows and margins of alpine lakes.
 ANIMAL FOOD: *Unspecified part*: eaten by all livestock, particularly horses, high value because of abundant fruit, U.S.A. (Hermann 1970).

var. **chalciolepis** (*Holm*) *F. J. Herm.*, Rhodora 70: 421 (1968).
Carex chalciolepis Holm, Amer. J. Sci. 4, 16: 28 (1903).
U.S.A.; rocky slopes, mountain meadows and alpine tundra.
 ANIMAL FOOD: *Unspecified part*: medium palatability and readily grazed by sheep and cattle, U.S.A. (Hermann 1970).

var. **brevisquama** *F. J. Herm*, Rhodora 70: 421 (1968).
N America, Greenland; mountain meadows, rocky arctic and alpine slopes.
 ANIMAL FOOD: *Unspecified part*: high in palatability, grazed by sheep, horses and, to a lesser extent, cattle, U.S.A. (Hermann 1970).

Carex hirta *L.*, Sp. Pl. 2: 975 (1753).
Europe, N Africa, temperate Asia; introduced into New Zealand; damp, sandy or grassy places, brackish swamps.
 WEEDS: *Perennial crops*: (Kühn 1982); *Grassland*: (Kühn 1982); *Waste places*: (Kühn 1982).

Carex hoodii *Boott* in Hook., Fl. Bor.-Amer. 2: 211 (1839).
N America; mountain meadows and slopes.
 ANIMAL FOOD: *Unspecified part*: high forage value for cattle and horses, N America (Hermann 1970).
 ENVIRONMENTAL USES: *Erosion control*: soil binder (Hermann 1970).

Carex humilis *Leyss.*, Fl. Halens.: 175 (1761).
Europe; dry places.
    ENVIRONMENTAL USES: *Ornamentals*: (Huxley 1992).

Carex idahoa *L. H. Bailey*, Bot. Gaz. (Crawfordsville) 21: 5 (1896).
U.S.A. (Montana and Idaho); mountain meadows.
    ANIMAL FOOD: *Unspecified part*: excellent for cattle and horses, U.S.A. (Hermann 1970).

Carex illota *L. H. Bailey*, Mem. Torrey Bot. Club 1: 15 (1889).
Canada and U.S.A.; high mountain meadows.
    ANIMAL FOOD: *Unspecified part*: important forage plant where abundant, U.S.A. (Hermann 1970).

Carex intumescens *Rudge*, Trans. Linn. Soc. London 7: 97, t. 9, f. 3 (1804).
Eastern N America; conifer forests, stream banks.
    ENVIRONMENTAL USES: *Ornamentals*: (Huxley 1992).

Carex inversa *R. Br.*, Prodr.: 242 (1810).
Australia and New Zealand; damp places, grasslands.
    ANIMAL FOOD: *Leaves/culms/aerial parts*: culm grazed (Anon. 1994 – 2000).
    WEEDS: *Gardens*: (Healy & Edgar 1980).

Carex inyx *Nelmes*, Proc. Linn. Soc. London 155: 279 (1944).
Australia and New Zealand; grasslands.
    WEEDS: *Grassland*: (Healy & Edgar 1980).

Carex jonesii *L. H. Bailey*, Mem. Torrey Bot. Club 1: 16 (1889).
U.S.A.; mountain meadows and along streams.
    ANIMAL FOOD: *Unspecified part*: fair to very good forage value for horses, cattle and sheep, U.S.A. (Hermann 1970).

Carex kaloides *Petrie*, Trans. & Proc. New Zealand Inst. 13: 332 (1881).
New Zealand; drier ground on margins of swamps, lakes and river margins.
    ENVIRONMENTAL USES: *Ornamentals*: (Huxley 1992).

Carex kelloggii *W. Boott* in S. Watson, Bot. California 2: 240 (1880).
Canada and U.S.A.; rocky lake margins, wet banks, moist to marshy meadows.
    ANIMAL FOOD: *Unspecified part*: commonly eaten by livestock when other feed is scarce, sheep thrive on it, U.S.A. (Hermann 1970).

Carex lanuginosa Michx., Fl. Bor.-Amer. 2: 175 (1803).
N America; wet meadows and stream banks.
    ANIMAL FOOD: *Unspecified part*: good forage value, palatable to all livestock, U.S.A. (Hermann 1970).

Carex lasiocarpa *Ehrh.*, Hannover. Mag. 9: 132 (1784).
Europe (rare in S), N Asia, N America; fens, bogs and lake margins.
WEEDS: *Aquatic biotopes*: (Kühn 1982).

Carex leporinella *Mack.*, Bull. Torrey Bot. Club 43: 605 (1917).
U.S.A.; rocky slopes, flats and meadows.
ANIMAL FOOD: *Unspecified part*: generally an infrequent species but of medium forage value to cattle where locally frequent, U.S.A. (Hermann 1970).

Carex longebrachiata *Boeck.*, Linnaea 41: 282 (1877).
Australia and New Zealand; grasslands.
WEEDS: *Grassland*: (Healy & Edgar 1980).

Carex luzulina *Olney*, Proc. Amer. Acad. Arts 7: 395 (1868).

var. **ablata** *(L. H. Bailey) F. J. Herm.*, Rhodora 70: 420 (1968).
Carex ablata L. H. Bailey, Bot. Gaz. (Crawfordsville) 13: 82 (1888), non Boott (1867).
C. luzulaefolia S. Watson var. *ablata* (L. H. Bailey) Kük. in Engl., Pflanzenr. 4 (20), 38 Heft: 558 (1909).
Canada and U.S.A.; mountain bogs and meadows.
ANIMAL FOOD: *Unspecified part*: good palatability, grazed by all livestock, U.S.A. (Hermann 1970).

Carex lyngbyei *Hornem.*, Nomencl. Fl. Danic.: t. 1888 (1827); Fors. Oecon. Plantel. ed. 3, 2: 275 (1837).
Circumpolar; swampy areas, wet meadows and margins of water bodies.
ANIMAL FOOD: *Unspecified part*: superb forage, Iceland (Ingvason 1969).

subsp. **cryptocarpa** (*C. A. Mey*) *Hultén*, Fl. Kamtchatka 1: 188 (1927).
Carex cryptocarpa C. A. Mey., Mém. Acad. Imp. Sci. St.-Pétersbourg, Divers Savans 1: 226 (1831).
Pacific coast of N Asia and N America; swampy areas, wet meadows and margins of water bodies.
ANIMAL FOOD: *Unspecified part*: highly rated forage for cattle and game animals, lends itself to silage-making, Russia (Ingvason 1969).
NOTE: Ingvason (1969) states that in Iceland the hybrids *Carex lyngbei* × *C. rigida* (= *C. bigelowii*) and *C. lyngbei* × *C. nigra* (L.) Reichard also make fair to good forage.

Carex macrocephala *Willd. ex Spreng.*, Syst. Veg. ed. 16, 3: 808 (1826).
N Asia, China, Japan; sandy areas, coastal.
MEDICINES: *Unspecified medicinal disorders*: China (Cauis & Banby 1935).

Carex maorica *Hamlin*, Trans. Roy. Soc. New Zealand 84: 684 (1957).
New Zealand; margins of water bodies.
WEEDS: *Aquatic biotopes*: (Healy & Edgar 1980); *Irrigation ditches*: (Healy & Edgar 1980).

Carex media *R. Br.* in Franklin, Narr. Journey Polar Sea: 763 (1823).
N America; moist, open or partially open habitats.
    ANIMAL FOOD: *Unspecified part*: grazed by cattle, sheep and horses, U.S.A. (Hermann 1970).

Carex mertensii *Prescott*, Mém. Acad. Imp. Sci. St.-Pétersbourg, Divers Savans 2: 168 (1832).
N America; rocky slopes and forests.
    ANIMAL FOOD: *Unspecified part*: good for cattle and horses, especially early in the season, closely grazed by sheep, U.S.A. (Hermann 1970).

Carex microptera *Mack.*, Muhlenbergia 5: 56 (1909).
N America; dry places, meadows and along streams in montane areas
    ANIMAL FOOD: *Unspecified part*: medium forage value for cattle and horses, low for sheep, U.S.A. (Hermann 1970).

Carex miserabilis *Mack.*, N. Amer. Fl. 18: 385 (1935).
U.S.A. (Washington, Oregon and Idaho); wet meadows, stream banks and lake margins.
    ANIMAL FOOD: *Unspecified part*: grazed by sheep, highly palatable to cattle, U.S.A. (Hermann 1970).

Carex montana *L.*, Sp. Pl. 2: 975 (1753).
Europe to C Russia; dry slopes, scrub and open woods.
    ENVIRONMENTAL USES: *Ornamentals*: (Huxley 1992).

Carex morrowii *Boott ex A. Gray*, Narr. Exped. Perry 2: 325 (1857).
C and S Japan; forests often by streams.
    ENVIRONMENTAL USES: *Ornamentals*: (Grounds 1989; Huxley 1992).

Carex multicostata *Mack.*, Bull. Torrey Bot. Club 43: 604 (1917).
U.S.A.; dry, often overgrazed meadows, and open montane woods.
    ANIMAL FOOD: *Unspecified part*: fair to good forage value for cattle and horses, U.S.A. (Hermann 1970).

Carex muskingumensis *Schwein.*, Ann. Lyceum Nat. Hist. New York 1: 66 (1824).
Western N America; wet swales.
    ENVIRONMENTAL USES: *Ornamentals*: (Grounds 1989; Huxley 1992).

Carex nebraskensis *Dewey*, Amer. J. Sci. Arts, ser. 2, 18: 102 (1854).
U.S.A. (particularly the Rockies), southern Canada; wet meadows, marshy or boggy ground, moist talus slopes.
    ANIMAL FOOD: *Unspecified part*: forage, U.S.A. (USDA-ARS 2000); medium potential for browsing, high potential for grazing, U.S.A. (USDA-NRCS 1999); an important forage plant for grazing and as a hay crop, valuable late-season sedge and then heavily grazed, U.S.A. (Hermann 1970).

Carex nelsonii *Mack.* in Rydb., Fl. Rocky Mts.: 137 (1917).
U.S.A. (Wyoming, Colorado, Utah); mountain meadows and rocky slopes.
    ANIMAL FOOD: *Unspecified part*: readily grazed by sheep, U.S.A. (Hermann 1970).

Carex nigra (*L.*) *Reichard*, Fl. Moeno-Francof. 2: 96 (1778).
Carex acuta L. var. *nigra* L., Sp. Pl. 2: 978 (1753).
Europe, W Asia, N America. Introduced to Australia; wet grasslands and marshy areas.
    ANIMAL FOOD: *Unspecified part*: good forage, Iceland, Russia (Ingvason 1969).
    ENVIRONMENTAL USES: *Ornamentals*: (Huxley 1992).

Carex nigricans *C. A. Mey.*, Mém. Acad. Imp. Sci. St.-Pétersbourg, Divers Savans 1: 211 (1831).
Western N America, NE Russia; alpine and spruce-fir meadows and tundra, often near melting snow.
    ANIMAL FOOD: *Unspecified part*: high palatability, heavily grazed by sheep and cattle, N America (Hermann 1970).

Carex nivalis *Boott*, Trans. Linn. Soc. London 20: 136 (1846).
Himalayas; open areas at high altitude.
    MEDICINES: *Injuries*: paste of powdered leaves applied as antiseptic on open wounds, India (Navchoo & Buth 1992).

Carex nova *L. H. Bailey*, J. Bot. 26: 322 (1888).
U.S.A.; mountain meadows and stream banks.
    ANIMAL FOOD: *Unspecified part*: excellent forage for all livestock, U.S.A. (Hermann 1970).

Carex obtusata *Lilj.*, Kongl. Vetensk. Acad. Nya Handl. 14: 69 (1793).
N America; dry plains, ridges and rocky open slopes.
    ANIMAL FOOD: *Unspecified part*: low forage producer but valuable for sheep and cattle especially in overgrazed areas, U.S.A. (Hermann 1970).

Carex oreocharis *Holm*, Amer. J. Sci. 4, 9: 358 (1900).
U.S.A.; dry slopes and grassland hills, granitic soil.
    ANIMAL FOOD: *Unspecified part*: grazed by cattle, medium palatability, U.S.A. (Hermann 1970).

Carex ornithopoda *Willd.*, Sp. Pl. 4: 255 (1805).
NW Europe; dry grassland and scrub.
    ENVIRONMENTAL USES: *Ornamentals*: (Grounds 1989; Huxley 1992).

Carex oshimensis *Nakai*, Bot. Mag. (Tokyo) 28: 326 (1914).
Japan; dry sandy and gravelly areas.
    ENVIRONMENTAL USES: *Ornamentals*: (Grounds 1989).

Carex ovalis *Gooden.*, Trans. Linn. Soc. London 2: 148 (1794).
Carex leporina auct. non L., Sp. Pl. 2: 973 (1753).
Europe, N America; damp grassy areas.
    WEEDS: *Waste places*: (Kühn 1982); *Grassland*: (Kühn 1982; Healy & Edgar 1980); *Irrigation ditches*: (Healy & Edgar 1980); *Forests*: (Healy & Edgar 1980).

Carex pachystachya *Cham. ex Steud.*, Syn. Pl. Glumac. 2: 197 (1855).
Canada, Aleutian Is., U.S.A.; meadows, open woods and slopes.
    ANIMAL FOOD: *Unspecified part*: good forage for all classes of livestock, U.S.A. (Hermann 1970).

Carex pachystylis *J. Gay*, Ann. Sci. Nat. Bot. sér. 2, 10: 301 (1838).
Temperate Asia, Egypt; dry, sandy, gravelly or rocky areas, dry pans.
    ANIMAL FOOD: *Unspecified part*: forage, Russia (USDA-ARS 2000).

Carex panicea *L.*, Sp. Pl. 2: 977 (1753).
Europe; marshes, heaths and damp grassland.
    WEEDS: *Rotation crops*: (Kühn 1982); *Perennial crops*: (Kühn 1982); *Grassland*: (Kühn 1982); *Aquatic biotopes*: (Kühn 1982).

Carex paniculata *L.*, Cent. Pl. 1: 32 (1755).
Europe, Russia; wet places, usually base-rich.
    MATERIALS: *Fibres*: used to make 'hassocks' (stools), U.K. (H283 — Fig. 1C); used to make brooms, U.K. (H295 — Fig. 2J); *Other materials/chemicals*: used in stables in place of straw (Heywood 1993; Huxley 1992).
    ENVIRONMENTAL USES: *Ornamentals*: (Heywood 1993; Huxley 1992).
    WEEDS: *Waste places*: (Kühn 1982); *Aquatic biotopes*: (Kühn 1982).

Carex pelocarpa *F. J. Herm.*, Rhodora 39: 492 (1937).
U.S.A.; alpine slopes, rocky lake shores, mountain meadows, stream beds.
    ANIMAL FOOD: *Unspecified part*: moderate to high forage value for cattle, sheep, goats and horses, U.S.A. (Hermann 1970).

Carex pendula *Huds.*, Fl. Angl.: 352 (1762).
Europe, W Asia, N Africa; woods and damp, shady places.
    ENVIRONMENTAL USES: *Ornamentals*: (Grounds 1989; Huxley 1992).

Carex petasata *Dewey*, Amer. J. Sci. 29: 246 (1836).
Canada and U.S.A.; rocky soil in mountain meadows and woods.
    ANIMAL FOOD: *Unspecified part*: good forage value for all livestock, U.S.A. (Hermann 1970).

Carex petriei *Cheeseman*, Trans. & Proc. New Zealand Inst. 16: 413 (1884).
New Zealand; montane stream banks, river flats and grasslands.
    ENVIRONMENTAL USES: *Ornamentals*: (Grounds 1989; Huxley 1992).

Carex phaeocephala *Piper*, Contr. U.S. Natl. Herb. 11: 172 (1906).
Canada and U.S.A.; high, rocky summits, occasionally coniferous woodland.
ANIMAL FOOD: *Unspecified part*: variable forage value for sheep, U.S.A. (Hermann 1970).

Carex phyllocephala *T. Koyama*, Acta Phytotax. Geobot. 16: 40 (1955).
Japan; habitat not recorded.
ENVIRONMENTAL USES: *Ornamentals*: (Grounds 1989).

Carex physodes *M. Bieb.*, Mém. Soc. Imp. Naturalistes Moscou 2: 104, t. 7 (1809).
Temperate Asia, European part of Russia; sandy, intermittently wet areas.
ANIMAL FOOD: *Unspecified part*: forage, Russia (USDA-ARS 2000).

Carex pilulifera *L.* Sp. Pl. 2: 976 (1753).
Europe, N Asia; mostly dry, grassy areas, rarely in open woods.
ENVIRONMENTAL USES: *Ornamentals*: (Grounds 1989).

FIG. 2. **A** *Schoenus nigricans* rope, Italy (EBMC 34392); **B** *Lepironia articulata* mat, China (EBMC 34372); **C** *Carex* sp. net, Australia (EBMC 34318); **D** *Lepidosperma squamatum* mat, Australia (EBMC 34370); **E** *Schoenoplectus lateriflorus* hat, Taiwan (EBMC 34405); **F** *Schoenoplectus lacustris* protective sleeve cuffs, origin unknown (EBMC 34404); **G** *Cyperus longus* paper, Channel Is. (EBMC 34357); **H** *Schoenoplectus lacustris* matting, U.K. (EBMC 34396); **J** *Carex paniculata* broom, U.K. (EBMC 34214); **K** *Cyperus longus* rope, Channel Is. (EBMC 34315). All from the Economic Botany Museum collections, Kew; catalogue number in brackets.

Carex pityophila *Mack.*, Bull. Torrey Bot. Club 40: 545 (1913).
U.S.A.; dry pine lands.
    ANIMAL FOOD: *Unspecified part*: moderately palatable to sheep and cattle, U.S.A. (Hermann 1970).

Carex plantaginea *Lam.*, Encycl. 3: 392 (1789).
N America; deciduous woodland.
    ENVIRONMENTAL USES: *Ornamentals*: (Grounds 1989; Huxley 1992).

Carex praegracilis *W. Boott*, Bot. Gaz. (Crawfordsville) 9: 87 (1884).
N America; moist open habitats on plains, occasionally montane.
    ANIMAL FOOD: *Unspecified part*: excellent forage value for horses and cattle, especially as winter grazing, N America (Hermann 1970).

Carex praticola *Rydb.,* Mem. New York Bot. Gard. 1: 84 (1900).
Canada and U.S.A.; high meadows, open woods, stream banks.
    ANIMAL FOOD: *Unspecified part*: moderate forage value for all livestock, especially sheep and goats, U.S.A. (Hermann 1970).

Carex preslii *Steud.*, Syn. Pl. Glumac. 2: 242 (1855).
Canada and U.S.A.; dryish, open slopes.
    ANIMAL FOOD: *Unspecified part*: good forage for cattle and horses, U.S.A. (Hermann 1970).

Carex pseudocyperus *L.*, Sp. Pl. 2: 978 (1753).
Europe, temperate Asia to Japan, N America; lake margins, ditches, fens.
    ENVIRONMENTAL USES: *Ornamentals*: (Huxley 1992).

Carex pseudoscirpoides *Rydb.*, Mem. New York Bot. Gard 1: 78 (1900).
U.S.A.; alpine tundra, rocky slopes, mountain meadows.
    ANIMAL FOOD: *Unspecified part*: moderately to highly palatable to sheep, cattle and horses, U.S.A. (Hermann 1970).

Carex pyrenaica *Wahlenb.*, Kongl. Vetensk. Acad. Nya Handl. 24: 139 (1803).
C and western N America, Eurasia; open rock or grassy alpine slopes.
    ANIMAL FOOD: *Unspecified part*: high palatability for cattle and horses, relatively low for sheep, N America (Hermann 1970).

Carex rariflora (*Wahlenb.*) *Sm.* in Sowerby, Engl. Bot.: t. 2516 (1813).
Carex limosa var. *rariflora* Wahlenb., Vetensk. Acad. Nya Handl. 24: 162 (1803).
Arctic, subarctic Eurasia and N America; bogs, marshes, salt marshes.
    ANIMAL FOOD: *Unspecified part*: good forage, Russia (Ingvason 1969).

Carex raynoldsii *Dewey*, Amer. J. Sci. 2, 32: 39 (1861).
Canada and U.S.A.; mountain meadows, open woods and rocky or grassy subalpine slopes.

ANIMAL FOOD: *Unspecified part*: palatability fair to good for cattle, sheep and horses, U.S.A. (Hermann 1970).

Carex rhynchophysa *Fisch., C. A. Mey. & Avé-Lall.*, Index Sem. Hort. Petrop. 9, Suppl.: 9 (1844).
Carex laevirostris Blytt ex Blytt & Fr., Bot. Not. 1844: 24 (1844).
Northern Eurasia; wet places.
MATERIALS: *Fibres*: used for making hats and table matting, Japan (H305, H306, H307).

Carex riparia *Curtis*, Fl. Londin. 2 (4): t. 60 (1783).
Europe to W Asia; wet grassy areas, ditches, marshes.
MATERIALS: *Other materials/chemicals*: used in stables in place of straw (Heywood 1993).
ENVIRONMENTAL USES: *Ornamentals*: (Grounds 1989; Huxley 1992).

Carex rossii *Boott* in Hook., Fl. Bor.-Amer. 2: 222 (1839).
N America; dry, open woods, in clearings, on slopes.
ANIMAL FOOD: *Unspecified part*: fair to good forage for sheep, U.S.A. (Hermann 1970).

Carex rostrata *Stokes* in With., Bot. Arr. Brit. Pl. ed. 2, 2: 1059 (1787).
N America, Greenland, Eurasia; wet meadows and marshes on lake margins.
ANIMAL FOOD: *Unspecified part*: good forage in spring, U.S.A. (Hermann 1970); excellent forage for cattle, Iceland (Ingvason 1969); rated very highly as aquatic forage, Russia (Ingvason 1969).

Carex rupestris *Bellardi ex All.*, Fl. Pedem. 2: 264, t. 92, f. 1 (1785).

var. **drummondii** (*Dewey*) *L. H. Bailey*, Cat. N. Amer. Caric. 4: (1884).
C. drummondiana Dewey, Amer. J. Sci. 29: 251 (1836).
N America; dry alpine summits and tundra.
ANIMAL FOOD: *Unspecified part*: low to moderate palatability, U.S.A. (Hermann 1970).

Carex saxatilis *L.*, Sp. Pl. 2: 976 (1753).

var. **major** *Olney* in S. Watson, Botany [fortieth parallel]: 370 (1871).
North America; rocky shores of mountain lakes and streams, wet mountain meadows.
ANIMAL FOOD: *Unspecified part*: low palatability, U.S.A. (Hermann 1970).

Carex scaposa *C. B. Clarke*, Bot. Mag. 113: t. 6940 (1887).
China; woods, scrubland, rocky areas, swamps.
ENVIRONMENTAL USES: *Ornamentals*: (Huxley 1992).

Carex scopulorum *Holm*, Amer. J. Sci 4, 14: 422 (1902).
U.S.A.; alpine meadows, lake shores, stream banks, tundra.

ANIMAL FOOD: *Unspecified part*: medium palatability to cattle, sheep and goats, U.S.A. (Hermann 1970).

var. **bracteosa** (*L. H. Bailey*) *F. J. Herm.*, Leafl. W. Bot. 9: 16 (1959).
Carex vulgaris Fr. var. *bracteosa* L. H. Bailey, Proc. Amer. Acad. Arts 22: 81 (1886).
Canada and U.S.A.; wet meadows.
    ANIMAL FOOD: *Unspecified part*: medium to high palatability, remaining green late in the season, U.S.A. (Hermann 1970).

Carex siderosticta *Hance*, J. Linn. Soc., Bot. 13: 89 (1873).
N Asia, China, Japan; shaded areas.
    MEDICINES: *Unspecified medicinal disorders*: India (Cauis & Banby 1935).
    ENVIRONMENTAL USES: *Ornamentals*: (Grounds 1989; Huxley 1992).

Carex simulata *Mack.*, Bull. Torrey Bot. Club 34: 604 (1908).
U.S.A.; wet meadows and swamps above 2500 m.
    ANIMAL FOOD: *Unspecified part*: moderate to high forage value, N America (Hermann 1970).

Carex spectabilis *Dewey*, Amer. J. Sci. 29: 248 (1836).
N America; mountain meadows and steep, rocky slopes.
    ANIMAL FOOD: *Unspecified part*: grazed by cattle, sheep and horses, U.S.A. (Hermann 1970).

Carex sprengelii *Dewey* in Spreng., Syst. Veg. 3: 827 (1826).
Canada and U.S.A.; Moist areas in loose gravel or sandy soil.
    ANIMAL FOOD: *Unspecified part*: fair to very good forage for all livestock, U.S.A. (Hermann 1970).

Carex subfusca *W. Boott* in S. Watson, Bot. California 2: 34 (1880).
Canada and U.S.A.; dry meadows and forest borders.
    ANIMAL FOOD: *Unspecified part*: medium forage value, U.S.A. (Hermann 1970).

Carex subnigricans *Stacey*, Leafl. W. Bot. 2: 167 (1939).
C and W U.S.A.; moist rocky slopes and alpine meadows.
    ANIMAL FOOD: *Unspecified part*: high palatability and heavily grazed by sheep, U.S.A. (Hermann 1970).

Carex sychnocephala *J. Carey*, Amer. J. Sci. 2, 4: 24 (1847).
Canada and U.S.A.; meadows, lake shores, thickets, gardens.
    ANIMAL FOOD: *Unspecified part*: grazed by livestock, Canada (Hermann 1970).
    WEEDS: *Gardens*: Canada, U.S.A. (Hermann 1970).

Carex sylvatica *Huds.*, Fl. Angl.: 353 (1762).
Europe, N Africa, temperate Asia, New Zealand; woods and other shaded areas.
    ENVIRONMENTAL USES: *Ornamentals*: (Huxley 1992).

Carex teneraeformis *Mack.*, Bull. Torrey Bot. Club 43: 609 (1917).
U.S.A.; meadows and open forests.
    ANIMAL FOOD: *Unspecified part*: good forage for cattle and horses but poor for sheep, U.S.A. (Hermann 1970).

Carex testacea *Sol. ex Boott* in Hook. f., Fl. Nov.-Zel. 1: 282 (1853).
New Zealand; forest, tussock grassland, sand-dunes.
    ENVIRONMENTAL USES: *Ornamentals*: (Huxley 1992).
    WEEDS: *Pasture*: (Parsons & Cuthbertson 1992).

Carex tolmiei *Boott* in Hook., Fl. Bor. Amer. 2: 224 (1839).
N America; mountain meadows and alpine slopes.
    ANIMAL FOOD: *Unspecified part*: regularly grazed by sheep, horses and, to a lesser extent, by cattle, resistant to heavy grazing, U.S.A. (Hermann 1970).
    ENVIRONMENTAL USES: *Erosion control*: important stabiliser of developing soils, U.S.A. (Hermann 1970).

Carex trifida *Cav.*, Icon. 5: 41, t. 465 (1799).
New Zealand; coastal areas.
    ENVIRONMENTAL USES: *Ornamentals*: (Huxley 1992).

Carex umbrosa *Host*, Icon. Descr. Gram. Austriac. 1: 52 (1801).
Europe, W temperate Asia; shaded areas, open, damp, stony or grassy slopes, bogs.
    ENVIRONMENTAL USES: *Ornamentals*: (Huxley 1992).

Carex uncifolia *Cheeseman*, Trans. & Proc. New Zealand Inst. 16: 412 (1884).
New Zealand; open montane areas.
    ENVIRONMENTAL USES: *Ornamentals*: (Grounds 1989; Huxley 1992).

Carex uruguiensis *Boeck.*, Bot. Jahrb. Syst. 7: 277 (1886).
Southern Brazil, Uruguay, northern Argentina; wet places.
    ENVIRONMENTAL USES: *Erosion control*: sand stabiliser, Brazil (Pio Corrêa 1926).

Carex vallicola *Dewey*, Amer. J. Sci. 2, 32: 40 (1861).
U.S.A.; dry open slopes and clearings.
    ANIMAL FOOD: *Unspecified part*: fair to excellent forage value, U.S.A. (Hermann 1970).

Carex vernacula *L. H. Bailey*, Bull. Torrey Bot. Club 20: 417 (1893).
C and W U.S.A.; open sunny slopes and stream banks from spruce-fir to alpine zones.
    ANIMAL FOOD: *Unspecified part*: U.S.A. (Hermann 1970).

Carex vesicaria *L.*, Sp. Pl. 2: 979 (1753).
N America, Eurasia; wet places.
    ANIMAL FOOD: *Unspecified part*: valuable forage especially for horses and cattle, occasionally sheep, U.S.A. (Hermann 1970).

Carex vulpina *L.*, Sp. Pl. 2: 973 (1753).
N Europe, Russia; wet or shaded areas.
ENVIRONMENTAL USES: *Ornamentals*: (Huxley 1992).

Carex vulpinoidea *Michx.*, Fl. Bor.-Amer. 2: 169 (1803).
N and S America, naturalised in Europe and New Zealand; damp to swampy areas.
WEEDS: *Waste places*: (Healy & Edgar 1980); *Pasture*: (Healy & Edgar 1980).

Carex xerantica *L. H. Bailey*, Bot. Gaz. (Crawfordsville) 17: 151 (1892).
Canada and U.S.A.; prairies and plains at mid-elevations, rarely montane.
ANIMAL FOOD: *Unspecified part*: low to moderate forage value, U.S.A. (Hermann 1970).

CLADIUM *P. Browne*

Cladium mariscus (*L.*) *Pohl*, Tent. Fl. Bohem. 1: 32 (1809).
Schoenus mariscus L., Sp. Pl. 1: 42 (1753).
Temperate and tropical regions worldwide; margins of lakes, pools, swamps and fens.
MATERIALS: *Fibres*: leaves used for tying material, U.S.A. (Hawaii), (Burkill 1985; Funk 1978; Heywood 1993); culms and leaves used for thatch, Europe and Africa (Burkill 1985; Heywood 1993; J. M. Lock pers. comm. 2001); culms used as a source of cheap paper (Heywood 1993).
ENVIRONMENTAL USES: *Ornamentals*: (Huxley 1992).
WEEDS: *Aquatic biotopes*: (Kühn 1982); *Rice fields*: (Kühn 1982).

subsp. **jamaicense** (*Crantz*) *Kük.*, Repert. Spec. Nov. Regni Veg. Beih. 40 (1): 523 (1938).
Cladium jamaicense Crantz, Inst. Rei Herb. 1: 362 (1766).
Widespread in America, Oceania, E Asia, Malesia; margins of lakes, pools, swamps and fens.
MATERIALS: *Fibres*: paper, Malay Peninsula, Americas (Burkill 1935; Burkill 1985).
ACTIVE COMPOUNDS PRESENT: *Alkaloid*: nutlets (Burkill 1985).

CAUSTIS *R. Br.*

Caustis dioica *R. Br.*, Prodr.: 239 (1810).
Australia; open, sandy or loamy soils.
ENVIRONMENTAL USES: *Revegetators:* mine waste revegetator, Australia (Rossetto *et al.* 1992); *Ornamentals*: cut and dried plant material, Australia (Rossetto *et al.* 1992).

COURTOISINA *Soják*

Courtoisina assimilis (*Steud.*) *P. Marquet*, Bull. Jard. Bot. Belg. 58 (1 – 2): 265 (1988).
Cyperus assimilis Steud., Flora 25: 584 (1842).
Courtoisia assimilis (Steud.) C. B. Clarke in T. Durand & Schinz, Consp. Fl. Afr. 5: 596 (1895).

Tropical Africa; seasonally wet areas, roadsides.

FOOD: *Rhizomes*: cooked and uncooked vegetable and famine food (Anon. 1994 – 2000); *Culms*: cooked vegetable and famine food (Anon. 1994 – 2000); *Leaves*: herb (Anon. 1994 – 2000).

Courtoisina cyperoides (*Roxb.*) *Soják*, Čas. Nár. Mus., Odd. Přír. 148 (3 – 4): 193 (1979, publ. 1980).
Kyllinga cyperoides Roxb., Fl. Ind. 1: 182 (1820).
Courtoisia cyperoides (Roxb.) Nees in Wight, Contr. Bot. India: 92 (1834).
Tropical E Africa, India to Thailand and Indo-China; seasonally wet places.

WEEDS: *Rice fields*: (Simpson pers. obs. 1994; Simpson & Koyama 1998).

CYMOPHYLLUS *Mack.*

Cymophyllus fraseri *Mack.* in Britton & Brown, Ill. Fl. N. U.S. 2 (1): 441 (1913).
Carex fraseri Andrews, Bot. Repos.: t. 639 (1811).
U.S.A.; montane areas.

ENVIRONMENTAL USES: *Ornamentals*: (Grounds 1989; Huxley 1992).

CYPERUS *L.*

Cyperus aggregatus (*Willd.*) *Endl.*, Cat. Hort. Acad. Vindob. 1: 93 (1842).
Mariscus aggregatus Willd., Enum. Hort. Berol. 70 (1809).
C and S America; wet places on sandy savannahs.

MEDICINES: *Unspecified medical disorders*: decoction of whole plant used by Tiriyó as external bath for fevers in children, Brazil (Milliken 1997).

Cyperus albostriatus *Schrad.*, Analecta Fl. Cap.: 7 (1832).
S Africa, adventive in Australia and New Zealand; grasslands, waste places.

ENVIRONMENTAL USES: *Ornamentals*: (Grounds 1989; Huxley 1992).

WEEDS: *Waste places*: (Healy & Edgar 1980); *Gardens*: (Healy & Edgar 1980).

Cyperus albosanguineus *Kük.* in Engl., Pflanzenr. 4(20), 101 Heft: 555 (1936).
Mariscus albosanguineus Napper, J. E. Africa Nat. Hist. Soc. Natl. Mus. 28: 16 (1971).
E tropical Africa; damp grasslands in upland areas.

ANIMAL FOOD: *Unspecified part*: grazed by all domestic livestock, Kenya (H271).

Cyperus alopecuroides *Rottb.*, Descr. Icon. Rar. Pl.: 38, t. 8, f. 2 (1773).
Africa (Mediterranean to the Tropics), Asia, N Australia, Caribbean; swamps and other wet habitats.

MATERIALS: *Fibres*: culms, matting and rush-bottomed chairs, cultivated, Egypt (Täckholm & Drar 1950); possibility of using culms for paper-making, Egypt (Täckholm & Drar 1950).

ENVIRONMENTAL USES: *Revegetators*: grown for the reclamation of saline land, Egypt (Täckholm & Drar 1950); *Ornamentals*: (Huxley 1992).

WEEDS: *Aquatic biotopes*: (Kühn 1982).

Cyperus alterniflorus *R. Br.*, Prodr.: 216 (1810).

Australia; margins of water bodies.

ANIMAL FOOD: *Unspecified part*: Australia (H156).

ENVIRONMENTAL USES: *Ornamentals*: (H156).

Cyperus alternifolius *L.*, Mant. Pl.: 38. (1767).

Cyperus involucratus Rottb., Descr. Pl. Rar.: 22 (1772).

C. flabelliformis Rottb., Descr. Icon. Rar. Pl.: 42, t. 12, f. 2 (1773).

C. alternifolius L. subsp. *flabelliformis* (Rottb.) Kük., Pflanzenr. 4 (20), 101 Heft: 193 (1936).

Tropical Africa; widely escaped from cultivation; waste places, river banks, damp, cultivated areas.

FOOD ADDITIVES: *Unspecified part*: dried and burnt, a solution of the ash used in cooking the leaves of other vegetables (Fox & Norwood Young 1982); *Rhizomes*: potash (Tredgold 1986).

ANIMAL FOOD: *Unspecified part*: all domestic livestock, Kenya (Burkill 1985).

MATERIALS: *Fibres*: culms and leaves used for mats and baskets (Tredgold 1986), E Africa (Greenway 1950), Ethiopia (H285 — Fig. 1F, H286 — Fig. 1G), Philippines (Burkill 1985; Kükenthal 1935 – 1936); used to make fish traps, Madagascar (H2); used to make fans and paintbrushes, Ethiopia (H327, H328).

SOCIAL USES: *'Religious' uses*: superstitious ritual, culms used as cure for sickness in humans and cattle, Ethiopia (Burkill 1985; Kükenthal 1935 – 1936); used in 'Atihena' ceremony to cover the dead body, Madgascar (H1).

MEDICINES: *Digestive system disorders*: rhizome administered to children for stomach ache, Tanzania (Burkill 1985); *Infections/infestations*: used for malarial fevers, Colombia (Milliken 1997); *Injuries*: dried plant ash applied to fresh wounds as disinfectant, Tanzania (Burkill 1985).

ENVIRONMENTAL USES: *Ornamentals*: (Burkill 1985; Huxley 1992; Kükenthal 1935 – 1936; Schermerhorn & Quimby 1960; Täckholm & Drar 1950; Tredgold 1986; H37).

WEEDS: *Waste places*: (Kühn 1982; H71); *Aquatic biotopes*: (Kühn 1982; Healy & Edgar 1980); *Rice fields*: (Kühn 1982; Simpson & Koyama 1998).

ACTIVE COMPOUNDS PRESENT: *Alkaloids*: present in rhizome and leaf (Burkill 1985, Kükenthal 1935 – 1936).

There is disagreement as to whether *C. alternifolius* and *C. involucratus* are distinct taxa. Some authors treat them as such while others regard *C. involucratus* as conspecific with *C. alternifolius*, sometimes being recognised as subsp. *flabelliformis*. In any case it is likely that many misidentifications have been made, particularly for material which has been introduced. We follow Kukkonen (1990) and place the taxa under *C. alternifolius*.

Cyperus amabilis *Vahl*, Enum. Pl. 2: 318 (1806).

Tropical Africa to India and Indo-China, tropical America; seasonally wet areas often in sandy soil.

ANIMAL FOOD: *Unspecified part*: cattle grazing, Senegal (Burkill 1985).

WEEDS: *Rotation crops*: (Kühn 1982); *Grassland*: (Kühn 1982); *Waste places*: (Kühn 1982).

Cyperus amauropus *Steud.*, Syn. Pl. Glumac. 2: 33 (1855).
Mariscus amauropus (Steud.) Cufod., Enum. Pl. Aeth. Sperm.: 1448 (1970).
E tropical Africa; grasslands, savannahs, shallow soils over rocks.
    FOOD: *Culms*: bulbous culm bases eaten by Masai, Kenya (H269).
    ANIMAL FOOD: *Unspecified part*: grazed by all domestic livestock, Kenya (H270).

Cyperus arenarius *Retz.*, Observ. Bot. 4: 9 (1786).
S and SW Asia; sandy areas.
    ANIMAL FOOD: *Unspecified part*: grazed by sheep, goats etc, India (Kulhari & Joshi 1992; H126).
    ENVIRONMENTAL USES: *Erosion control*: live plant drought resistant, used in dune stabilisation, India (Anon. 1994 – 2000; Mathur & Govil 1987).
    WEEDS: *Unspecified weed*: Bahrain (H125).

Cyperus articulatus *L.*, Sp. Pl. 1: 44 (1753).
Pantropical; sometimes cultivated; swamps and other wet places, often in standing water.
    FOOD: *Unspecified part*: vegetable (Anon. 1994 – 2000); *Rhizomes*: fresh rhizomes peeled and eaten raw, sometimes grown and stored until needed (Anon. 1994 – 2000; Täckholm & Drar 1950).
    ANIMAL FOOD: *Unspecified part*: eaten by cattle when nothing else available (Täckholm & Drar 1950); eaten by cattle, goats and sheep (H243).
    MATERIALS: *Unspecified material type*: (H112); perfumery (Anon. 1994 – 2000); *Fibres*: culms used for matting (Anon. 1994 – 2000), Ghana, Nigeria, Sierra Leone (Burkill 1985; Täckholm & Drar 1950; H111, H115, H179, H181), Tanzania (H245); used for mattresses and sacking, W Africa (Abbiw 1990; Burkill 1985); beaten flat and woven into baskets, Ghana (Abbiw 1990), Tanzania (H244); used in square-bottomed fish baskets, Ghana (Burkill 1985); *Other materials/chemicals*: rhizomes burnt for incense, fresh rhizomes rubbed on rims of water jars for fragrance, also dried ash, dried rhizomes tied to hair for perfume, dried rhizomes or ash kept in cloth-boxes to fragrance clothes; fresh rhizomes put in jars to clarify muddy water, Africa (Abbiw 1990; Burkill 1985; Heywood 1993; Täckholm & Drar 1950; H110, H114); fragrant tuber sold under the name of 'kajiji' (Hausa), Nigeria (H114, H180); cultivated for its fragrant rhizome, Sierra Leone (H112, H177, H178); yields essential oil useful in the perfumery industry, Ghana (Abbiw 1990; Heywood 1993); rubbed on the body as a scent, Sierra Leone (H110); body perfume and cleanser, mixed with oil or soap (Burkill 1985); suggested cultivation for commercial soap-making (Burkill 1985); cultivated for aromatic rhizome, dried and powdered, as a fumigant, mixed with scented resins for clothes and to sweeten air of rooms in rainy season, W Africa (Burkill 1985); rhizomes burnt over fire, used as mosquito repellant, Lake Chad (Burkill 1985); made into necklaces and waist girdles for use as insect repellant (Burkill 1985).
    SOCIAL USES: '*Religious*' *uses*: rhizome used to fumigate the body during sickness,

Nigeria (Burkill 1985); rhizomes used to counteract witchcraft, Ecuador (Plowman *et al.* 1990); rhizome grated, mixed with water and given as a drink to children to increase intelligence, Venezuela (Plowman *et al.* 1990); rhizome chewed by defendant in court as a charm to secure acquittal, Nigeria (Burkill 1985); used as a charm to ensure success in trade and to exorcise evil spirits causing disease (Burkill 1985); part of culm worn by mother of twins on body to protect her and babies, Gabon (Burkill 1985); rhizome worn as a necklace by mother after childbirth, and chewed by her to quieten a restless child, Tanzania (Burkill 1985); commonly grown in sacred groves, the scented rhizome being offered to the dead (Burkill 1985); plant debris sprinkled by witch doctors over the body of patient to give them strength (Burkill 1985); Shipibo-Conibo fisherman rub the rhizome over the body before fishing, Peru (Tournon *et al.* 1986); placed on the nose of men who do not want to work, Shipibo-Conibo, Peru (Tournon *et al.* 1986); rhizome of plant parasitised by the fungus *Cintractia limitata* G. P. Clinton (*Ustilaginaceae*) is rubbed on to machetes for better work in the fields, Shipibo-Conibo, Peru (Tournon *et al.* 1986); arrows rubbed with the rhizome will reach animals being hunted, Shipibo-Conibo, Peru (Tournon *et al.* 1986); to reconcile couples the rhizomes are grated and placed in food or rubbed on clothes or the bed, Shipibo-Conibo, Peru (Tournon *et al.* 1986); *Miscellaneous social uses*: rhizome of plant parasitised with *Cintractia axicola* (Berk.) Cornu f. *peribebuyensis* (Sawada) Zambett. (*Ustilaginaceae*) is used to wash and preserve the hair, Shipibo-Conibo, Peru (Tournon *et al.* 1986); used as an aphrodisiac, Shipibo-Conibo, Peru (Tournon *et al.* 1986).

MEDICINES: *Unspecified medicinal disorders*: tuber used as tonic and stimulant, W Africa (Cauis & Banby 1935); important medicinal plants in culture of Shuar and Achuar, Ecuador (Evans 1991); given to children to make them grow taller and prevent them falling ill, Shipibo-Conibo, Peru (Tournon *et al.* 1986); crushed rhizome of plant parasitised with the fungus *Cintractia axicola* f. *peribebuyensis* applied to traumatised parts of body, Shipibo-Conibo, Peru (Tournon *et al.* 1986); *Circulatory system disorders*: rhizomes an ingredient of a medicinal drink used in the treatment of haemorrhoids (Kambu *et al.* 1989); *Digestive system disorders*: decoction of rhizome drunk as remedy against colic, Egypt (Täckholm & Drar 1950); given to small children with diarrhoea, Colombia (Plowman *et al.* 1990); rhizomes an ingredient of a medicinal drink used in the treatment of constipation, diarrhoea and gastritis (Kambu *et al.* 1989); rhizome of plant parasitised with the fungus *Cintractia limitata* (*Ustilaginaceae*) is crushed in water and taken in cases of diarrhoea, Shipibo-Conibo, Peru (Tournon *et al.* 1986); *Genito-urinary system disorders*: beneficial in menstrual disorders, Senegal, Congo (Burkill 1985); culm used for the treatment of menorrhagia: old culm burnt with whole fruit of *Aframomum melegueta* K. Schum. (*Zingiberaceae*) and swallowed with cold water, Nigeria (Adjanohoun *et al.* 1991); rhizomes an ingredient of medicinal drink used in treatment of sexual weakness, Congo (Kinshasa) (Kambu *et al.* 1989); *Ill defined symptoms*: rhizomes an ingredient of medicinal drink used in treatment of weakness (Kambu *et al.* 1989); to calm quick-tempered people, the rhizome is crushed in water and the water drunk, with a little also added to food, Shipibo-Conibo, Peru (Tournon *et al.* 1986); *Infections/infestations*: vermicidal, Congo, Tanzania (Burkill 1985); used to treat worms (Kükenthal 1935 – 1936); decoction of rhizome drunk as an antimalarial, Tanzania (Burkill 1985); antiemetic and sedative, used to stop vomiting in 'yellow fever' (Burkill 1985); pulped rhizomes rubbed onto bodies of babies with fever, Congo (Burkill 1985); used in treatment of endemic fever, W Indies (H313); *Inflammation*: rhizome used in poultice applied to swollen areas (Burkill 1985); *Mental disorders*: rhizome used as a sedative (Anon. 1994 – 2000); *Muscular-skeletal system disorders*: rhizomes an ingredient of medicinal drink used in treatment of back pain (Kambu *et al.* 1989); *Pain*: rhizome powdered and applied to head or used as an inhalant for migraine, Gabon, Senegal, Nigeria (Abbiw 1990; Burkill 1985); rhizome used as a toothache remedy, East Africa (Burkill 1985); *Poisonings*: juice of rhizome used for snakebite, Ecuador (Plowman *et al.* 1990); *Pregnancy/birth/puerpuerium disorders*: rhizome used as a dressing on the umbilical cord by the Shipibo-Conibo, Peru (Tournon *et al.* 1986); *Respiratory system disorders*: decoction drunk for all respiratory problems, Congo (Burkill 1985); rhizome chewed for coughs Nigeria, Ghana (Abbiw 1990; Burkill 1985); *Sensory system disoroders*: crushed rhizome applied to the eyes to improve sight, Shipibo-Conibo, Peru (Tournon *et al.* 1986); *Skin/subcutaneous cellular tissue disorders*: pulped rhizome rubbed into epidermal scarifications for oedoma and rheumatism, Congo (Burkill 1985); tuber ground up and applied to make hair grow thicker, Ecuador (Plowman *et al.* 1990); rhizomes powdered with clay and applied to skin disorders Nigeria, Sierra Leone (Burkill 1985; H110, H148).

ENVIRONMENTAL USES: *Erosion control*: planted on channels of sandy fields for retaining banks (Täckholm & Drar 1950).

WEEDS: *Unspecified weed*: (Abbiw 1990; H113); *Rotation crops*: (Kühn 1982); *Perennial crops*: (Kühn 1982); *Aquatic biotopes*: Ghana (Abbiw 1990; Kühn 1982); Sierra Leone (H113).

Cyperus aucheri *Jaub. & Spach.*, Ill. Pl. Orient. 2, 1: t. 101 (1844).
N Africa, Arabia to Iraq and Iran; sandy areas.

FOOD: *Culms*: culms edible and sweet, United Arab Emirates (H131).

ANIMAL FOOD: *Unspecified part*: grazed by camels, sheep and probably goats, Iraq, Kuwait, Oman, Saudi Arabia (H127, H128, H129, H130, H132).

MATERIALS: *Fibres*: culms used for making ropes, Kuwait (H130).

Cyperus bifax *C. B. Clarke*, Kew Bull. Addit. Ser. 8: 12 (1908).
Australia; wet or flooded areas.

MEDICINES: *Infections/infestations*: culms used for treatment of gonorrhoea, Australia (Anon. 1994 – 2000).

Cyperus boreohemisphaericus *Lye*, Lidia 3 (3): 77 (1993).
Ethiopia; grassy places, montane forest.

WEEDS: *Rotation crops*: Ethiopia (H234); *Perennial crops*: Ethiopia (H235).

Cyperus bulbosus *Vahl*, Enum. Pl. 2: 342 (1806).
Old World tropics; seasonally wet grasslands.

FOOD: *Tubers*: edible, roasted or boiled, Africa, Australia, Sri Lanka (Anon. 1994 – 2000; Kükenthal 1935 – 1936; H66, H311); famine food (Burkill 1935; Täckholm

& Drar 1950; Wickens *et al.* 1989); tuber fed to infants, Burkina Faso, Somalia (Burkill 1985).

ANIMAL FOOD: *Unspecified part*: grazing, Sudan, Somalia (Burkill 1985), Australia (H66).

ENVIRONMENTAL USES: *Ornamentals*: (H66).

WEEDS: *Cultivation*: Sudan, Tanzania (Burkill 1985); *Rotation crops*: Kenya (H257).

Cyperus camphoratus *Liebm.*, Kongel. Danske Vidensk. Selsk. Skr., Naturvidensk Math. Afd., ser. 5, 2: 216 (1851).
Tropical America; habitat not recorded.

SOCIAL USES: *Miscellaneous social uses*: culm bases used as an aphrodisiac (Kükenthal 1935 – 1936).

MEDICINES: *Digestive system disorders*: culm base enhances digestion (Kükenthal 1935 – 1936); *Infections/infestations*: culm base a febrifuge (Kükenthal 1935 – 1936); *Metabolic system disorders*: culm base induces sweating (Kükenthal 1935 – 1936); *Pregnancy/birth/puerperium disorders*: culm base used to assist labour (Kükenthal 1935 – 1936).

Cyperus canus *C. Presl* in J. Presl & C. Presl, Reliq. Haenk. 1: 179 (1828).
Tropical America; margins of water bodies and in water.

MATERIALS: *Fibres*: leaves used for grass matting, coasters and fans (Anon. 1994 – 2000); split culms used for mats and baskets, Mexico (Kükenthal 1935 – 1936).

Cyperus capitatus *Vand.*, Fasc. Pl.: 5 (1771).
Mediterranean, Canary Is; coastal sandy areas.

ANIMAL FOOD: *Unspecified part*: Turkey (H123).

ENVIRONMENTAL USES: *Erosion control*: potential sand binder, Egypt (H124).

Cyperus castaneus *Willd.*, Sp. Pl. 1: 278 (1797).
India to northern Australia; open, sandy and seasonally wet areas.

ANIMAL FOOD: *Unspecified part*: grazed by buffalo, Malaysia (Burkill 1935).

Cyperus chordorrhizus *Chiov.*, Agric. Colon. 20: 105 (1926).
Somalia, Kenya; coastal sand-dunes.

ANIMAL FOOD: *Unspecified part*: moderately grazed, Somalia (H229).

ENVIRONMENTAL USES: *Erosion control*: sand stabiliser (Fagotto 1987); main sand-binding plant in coastal belt, Somalia (H228); *Revegetators*: common pioneer of cemented sands, Somalia (Fagotto 1987).

Cyperus clarus *S. T. Blake*, Proc. Roy. Soc. Queensland 51: 44 (1939).
Australia; muddy depressions and hollows, margins of cultivated ground.

ANIMAL FOOD: *Leaves/culms/aerial parts*: leaves, Australia (Anon. 1994 – 2000).

Cyperus commixtus *Kük.*, Repert. Spec. Nov. Regni Veg. 29: 195 (1931).
Somalia; pool margins, marshy ground.

FOOD: *Aerial parts*: young vegetative apices, Somalia (Kükenthal 1935 – 1936).

Cyperus compactus *Retz.*, Observ. Bot. 5: 10 (1789).
Widespread in India and SE Asia, introduced into the Mascarenes; swampy areas, rice-fields, ditches and river banks.

MATERIALS: *Fibres*: sometimes used to roof houses, Thailand (H101).

WEEDS: *Cultivation*: (Simpson 1992a); *Rotation crops*: soya bean fields, Thailand (Radanachaless & Maxwell 1994); *Perennial crops*: (Kühn 1982); *Waste places*: (Kühn 1982); *Aquatic biotopes*: (Kühn 1982); *Rice fields*: (Kühn 1982; Simpson & Koyama 1998).

Cyperus compressus *L.*, Sp. Pl. 1: 46 (1753).
Pantropical; common; grasslands, waste places, cultivated areas.

FOOD: *Rhizomes*: cooked and uncooked vegetable (Anon. 1994 – 2000); famine food (Anon. 1994 – 2000).

ANIMAL FOOD: *Unspecified part*: grazed cattle and buffalo, SE Asia (Burkill 1935; Burkill 1985; H898); grazed by asses but poor value, India (Kulhari & Joshi 1992); forage, Brazil (Pio Corrêa 1926).

MATERIALS: *Other materials/chemicals*: rhizomes used for scenting oil, Tonga (Altschul 1973).

MEDICINES: *Infections/infestations*: roasted tubers made into paste and mixed with coconut oil for killing lice, India (Deokule & Magdum 1992).

ENVIRONMENTAL USES: *Ornamentals*: (Huxley 1992).

WEEDS: *Cultivation*: (Burkill 1935; Burkill 1985; Dangol 1992; H35, H36); *Rotation crops*: (Kühn 1982); soya bean fields, Thailand (Radanachaless & Maxwell 1994); *Grassland*: Nigeria (Kühn 1982; H172); *Waste places*: (Burkill 1985; Kühn 1982; H51, H58, H168, H169, H170, H171).

Cyperus congestus *Vahl*, Enum. Pl. 2: 358 (1806).
Southern Africa, introduced into Australia and New Zealand; waste places, damp areas, particularly margins of water bodies.

FOOD: *Unspecified part*: made into a sauce for vegetables (Kükenthal 1935 – 1936); *Tubers*: (Fox & Norwood Young 1982; Peters 1992); light roasting improves flavour, an important food of Kung Bushmen, Namibia (Fox & Norwood Young 1982).

ENVIRONMENTAL USES: *Ornamentals*: (Huxley 1992).

WEEDS: *Waste places*: weed of roadsides (Healy & Edgar 1980); *Aquatic biotopes*: weed on margins of rivers streams, swamps, roadside gutters and ditches (Healy & Edgar 1980); *Gardens*: (Healy & Edgar 1980).

Cyperus conglomeratus *Rottb.*, Descr. Icon. Rar. Pl.: 21, t. 15, f. 7 (1773).
Africa, Arabia to India; sandy soils, clay soils, rocky areas.

ANIMAL FOOD: *Unspecified part*: grazing for camels, goats and sheep livestock (Al-Zoghet 1989; Anon. 1994 – 2000; Burkill 1985).

MEDICINES: *Digestive system disorders*: used to treat diarrhoea and dysentery, Saudi Arabia (Al-Zoghet 1989); *Infections/infestations*: used to treat intestinal parasites, Saudi Arabia (Al-Zoghet 1989).

ENVIRONMENTAL USES: *Unspecified environmental uses*: marginal use for landscaping, Saudi Arabia (Al-Zoghet 1989); *Erosion control*: sand-binder, dune stabilisation, ground cover, Saudi Arabia, Africa (Al-Zoghet 1989; Burkill 1985).

WEEDS: *Cultivation*: (H100).

var. **effusus** (*Rottb.*) *Kük.* in Engl., Pflanzenr. 4 (20), 101 Heft: 275 (1935).
C. effusus Rottb., Descr. Icon. Rar. Pl.: 22, t. 12, f. 3 (1773).
Middle East to India; sandy ground, often near coast.
    ANIMAL FOOD: *Unspecified part*: India (H142).

Cyperus corymbosus *Rottb.*, Descr. Icon. Rar. Pl.: 42, t. 7, f. 4 (1773).
Tropical Africa, Madagascar, India to Indo-China, northern Australia, West Indies
and tropical S America; cultivated in N Africa, Asia; marshy places.
    MATERIALS: *Unspecified material type*: Algeria, Taiwan (H136); *Fibres*: culms
cultivated, dried and used for basketry, hats, matting and ropework, Fig. 3B —
H298, Borneo, India, China, Sri Lanka, Taiwan, Thailand (Simpson 1992a; H137,
H138, H139, H141, H298, H308, H316, H318).

var. **longispiculatus** (*Kuntze*) *Kük.* in Engl., Pflanzenr. 4 (20), 101 Heft: 82 (1935).
Cyperus enodis Boeck. var. *longispiculatus* Kuntze, Revis. Gen. Pl. 2: 749 (1891).
India, Indo-China, S China, N Australia; marshy places.
    MATERIALS: *Fibres*: matting and string (Burkill 1935; Heywood 1993;
Schermerhorn & Quimby 1960).

var. **scariosus** (*R. Br.*) *Kük.* in Engl., Pflanzenr. 4 (20), 101 Heft: 83 (1935).
Cyperus scariosus R. Br., Prodr.: 216 (1810).
India to N Australia; coastal, swampy brackish areas.
    MATERIALS: *Other materials/chemicals*: sweet-smelling rhizomes used in perfumery
and cosmetics (Kükenthal 1935 – 1936).
    MEDICINES: *Unspecified medicinal disorders*: useful in treatment of chest disorders and
nasal discharge; blood enricher, India (Cauis & Banby 1935); *Circulatory system disorders*:
useful in treatment of haemorrhoids, India (Cauis & Banby 1935); *Digestive system
disorders*: useful in the treatment of biliousness, thirst relief, fatigue, flatulence,
diarrhoea, India (Cauis & Banby 1935); *Genito-urinary system disorders*: rhizome diuretic,
stimulates menstrual discharge, also checks abnormally profuse menstruation and
urination, India (Cauis & Banby 1935); *Infections/infestations*: useful in the treatment of
fever and dysentry, India (Cauis & Banby 1935); decoction used to treat gonorrhoea
and syphilis, India (Cauis & Banby 1935); *Inflammation*: useful in treatment of
swellings and eye sores, India (Cauis & Banby 1935); *Metabolic system disorders*: rhizome
pungent, acrid and cooling, promoting the flow of milk (Cauis & Banby 1935);
regulates body temperature by antipyretic and sedative action, India (Cauis & Banby
1935); rhizome considered to be diaphoretic, India (Cauis & Banby 1935); *Nervous
system disorders*: useful in treatment of epilepsy, stuttering, brain disorders, India (Cauis
& Banby 1935); *Pain*: useful in treatment of lumbago, India (Cauis & Banby 1935);
Poisonings: useful in treatment of scorpion sting, India (Cauis & Banby 1935).
    WEEDS: *Unspecified weed*: Australia (H102).

Cyperus crassipes *Vahl*, Enum. Pl. 2: 299 (1806).
Tropical Africa; dry sandy places both coastal and inland.

ENVIRONMENTAL USES: *Erosion control*: sand-binder (Burkill 1985).

Cyperus cuspidatus *Kunth* in Humb., Bonpl. & Kunth, Nov. Gen. Sp. 1: 204 (1816).
Pantropical; open, often sandy areas.
ANIMAL FOOD: *Unspecified part*: Senegal (H215).
WEEDS: *Rotation crops*: (Kühn 1982); Nigeria (H214); maize fields (Dangol 1992); soya bean fields, Thailand (Radanachaless & Maxwell 1994); *Waste places*: (Kühn 1982); *Aquatic biotopes*: (Kühn 1982); *Rice fields*: (Simpson & Koyama 1998).

Cyperus cyperinus (*Retz.*) *J. V. Suringar*, Cyperus: 154, t. 6, f. 10 (1898).
Kyllinga cyperina Retz., Observ. Bot. 6: 21 (1791).
Tropics from India to N Australia and Polynesia; open, wet to seasonally wet areas.
WEEDS: *Unspecified weed*: (H144, H146); *Cultivation*: (H143, H145); *Gardens*: (H147); *Rice fields*: (Simpson & Koyama 1998).

Cyperus cyperoides (*L.*) *Kuntze*, Revis. Gen. Pl. 3 (2): 333 (1898).
Scirpus cyperoides L., Mant. Pl. 2: 181. 1771.
Old World Tropics, W Indies; open or slightly shaded areas.
FOOD: *Tubers*: used as famine food, Ghana (Abbiw 1990).
VERTEBRATE POISONS: *Mammals*: used as a vermifuge, Sumatra (Heywood 1993).
MEDICINES: *Injuries*: plant ash used to heal wounds, Nepal (Manandhar 1989); *Pain*: infusion of nutlets used for toothache, Philippines (Altschul 1973; Millar & Morris 1988); *Skin/ subcutaneous cellular tissue disorders*: ground rhizome applied to skin disorders, Ghana (Abbiw 1990).
ENVIRONMENTAL USES: *Ornamentals*: (Huxley 1992).
WEEDS: *Unspecified weed*: (H106); *Cultivation*: (Dangol 1992; H60, H149); *Rotation crops* (Kühn 1982); Congo (Kinshasa) (H224); *Perennial crops*: (Kühn 1982); *Grassland*: (Kühn 1982); *Waste places*: (Burkill 1985; Kühn 1982); *Disturbed land*: roadside weed (H105); *Rice fields*: (Simpson & Koyama 1998).

Cyperus denudatus *L. f.*, Suppl.: 102 (1782).
C E and S tropical Africa, Madagascar; swamps, swamp margins, ditches and other wet habitats.
WEEDS: *Rice fields*: (H260)

Cyperus dichroöstachyus *A. Rich.*, Tent. Fl. Abyss. 2: 481 (1851).
Tropical Africa; swamps, wet areas by streams.
ANIMAL FOOD: *Unspecified part*: grazed by all domestic livestock, Kenya (Burkill 1985).
MATERIALS: *Fibres*: thatching, Kenya (Burkill 1985).

Cyperus difformis *L.*, Cent. Pl. 2: 6 (1756).
Pantropical; also subtropical and warm temperate regions; wet and swampy areas.
ANIMAL FOOD: *Unspecified part*: grazing, domestic livestock, Kenya (Burkill 1985; H25).
SOCIAL USES: *Miscellaneous social uses*: crushed rhizome used as an aphrodisiac by

the Shipibo-Conibo, Peru (Tournon *et al.* 1986).

WEEDS: *Unspecified weed*: (Abbiw 1990; Burkill 1985; H25); *Cultivation*: (Burkill 1985; Dangol 1992; Kühn 1982; Täckholm & Drar 1950; H26, H54); *Rotation crops*: soya bean fields, Thailand (Radanachaless & Maxwell 1994); *Grassland*: (Kühn 1982); *Aquatic biotopes*: (Kühn 1982); *Rice fields:* (Burkill 1985; Dangol 1992; Kühn 1982; Simpson & Koyama 1998; Täckholm & Drar 1950; H26, H54).

Cyperus digitatus *Roxb.* in Carey, Fl. Ind. 1: 209 (1820).
Pantropical; wet and swampy areas.

MATERIALS: *Fibres*: matting, Java (Burkill 1935); poor quality roofing, Thailand (H46); culms used for weaving mats and baskets (Tredgold 1986); *Other materials/chemicals*: potash (Tredgold 1986).

MEDICINES: *Respiratory system disorders*: leaves used to treat coughs, Uganda (Adjanohoun *et al.* 1993).

WEEDS: *Cultivation*: (Kühn 1982); *Perennial crops*: Somalia (H226); *Waste places*: (Kühn 1982); *Aquatic biotopes*: (Kühn 1982); *Rice fields*: (Kühn 1982; Simpson & Koyama 1998).

subsp. **auricomus** (*Sieber ex Spreng.*) *Kük.*, Bot. Not. 1934: 65 (1934).
C. auricomus Sieber ex Spreng., Syst. Veg. 1: 230 (1825).
Tropical Africa; swampy areas or usually near water.

MATERIALS: *Other materials/chemicals*: used as substitute for perfume, Sudan (Burkill 1985), Nigeria (H176).

WEEDS: *Unspecified weed*: Nigeria (H109); *Cultivation*: (Burkill 1985; H108).

Cyperus dilatatus *Schum. & Thonn.*, Beskr. Guin. Pl.: 38 (1827).
Cyperus gracilinux C. B. Clarke, J. Linn. Soc., Bot. 21: 162 (1881); in Dyer, Fl. Trop. Afr. 8: 362 (1902).
Tropical Africa; seasonally wet habitats.

MEDICINES: *Infections/infestations*: rhizome used for treatment of gonorrhoea, Nigeria (Burkill 1985); *Injuries*: culm used for healing cuts and wounds, Nigeria (Burkill 1985).

WEEDS: *Unspecified weed*: (H159); *Cultivation*: Ivory Coast (H189); *Grassland*: (H157, H160); *Disturbed land*: (H162); *Gardens*: Nigeria, Gambia (Burkill 1985; H91, H158, H161).

Cyperus distans *L. f.*, Suppl.: 103 (1782).
Pantropical; damp grassland, areas of cultivation, stream banks, roadsides.

FOOD: *Rhizomes*: rhizome used in a sauce, Ghana (Abbiw 1990), Guinea (Burkill 1985; H49).

FOOD ADDITIVES: *Unspecified part*: ashes extracted for salt, Zambia (H278).

ANIMAL FOOD: *Unspecified part*: Indonesia, Sabah (H49); grazed by cattle, sheep and goats, Kenya (H258).

MATERIALS: *Other materials/chemicals*: rhizome used for scenting coconut oil, Fiji (H78).

MEDICINES: *Infections/infestations*: rhizome used as treatment for gonorrhoea,

Nigeria (Burkill 1985); *Injuries*: culm used for healing cuts and wounds, Nigeria (Burkill 1985); *Pain*: infusion of leaves used to treat chest pain, Uganda (Adjanohoun *et al.* 1993).

WEEDS: *Unspecified weed*: (Abbiw 1990; H57); *Cultivation*: Ivory Coast, India (Burkill 1985; Dangol 1992; H166), Gambia (Kühn 1982; H165), Zambia (H277); *Perennial crops*: (Kühn 1982); *Grassland* (Kühn 1982); *Waste places*: (Kühn 1982); *Aquatic biotopes*: (Kühn 1982); *Gardens*: (H42, H43, H44).

Cyperus dives *Delile*, Fl. Aegypt. Illustr.: 5, t. 4, f. 3 (1813).
Tropical Africa, Middle-East to India; swampy areas, river banks.
ANIMAL FOOD: *Unspecified part*: Ethiopia (H279), Botswana (H273).
MATERIALS: *Fibres*: thatching, Kenya (H242); culms used for basket making, Uganda (Burkill 1985); culms used for matting, Egypt (Täckholm & Drar 1950).

Cyperus dubius Rottb., Descr. Icon. Rar. Pl.: 20, t. 4, f. 5. 1773.
Mariscus dubius (Rottb.) Kük. ex Fisch. in Gamble, Fl. Madras 9: 1644 (1931).
Tropical Africa to India, Indo-China and Malesia; sandy foreshores and rock crevices or shallow soils inland.
MATERIALS: *Other materials/chemicals*: rhizome aromatic with a strong sweet scent, Ghana (H207, H209).
WEEDS: *Rotation crops*: Ivory coast (H208); *Grassland*: Ghana (H209).

Cyperus elatus *L.*, Cent. Pl. 2: 301 (1756).
India to W Malesia; swamps, river-banks and rice-fields.
ANIMAL FOOD: *Unspecified part*: cattle fodder (Burkill 1935).
MATERIALS: *Fibres*: culms used for hats, Java, Sulawesi (Kern 1974); sleeping mats, Vietnam, Java, Sulawesi (Burkill 1935).
WEEDS: *Rice fields*: (Simpson & Koyama 1998).

Cyperus elegans *L.*, Sp. Pl. 1: 45 (1753).
Southern N America, C America, W Indies; dryish to wet sandy soils.
MATERIALS: *Other materials/chemicals*: sweet-smelling rhizomes used in perfumery and cosmetics (Kükenthal 1935 – 1936).
MEDICINES: *Unspecified medicinal disorders*: India (Cauis & Banby 1935).

Cyperus entrerianus *Boeck.*, Flora 61: 139 (1878).
S America (Argentina, Brazil, Paraguay, Uruguay), naturalised in southern N America; open grasslands.
WEEDS: *Unspecified weed*: U.S.A. (USDA-ARS 2000).

Cyperus eragrostis *Lam.*, Tabl. Encycl. 1: 146 (1791).
N and S America, introduced into Europe, parts of SE Asia, Australia and New Zealand; damp and wet places on margins of water bodies.
ANIMAL FOOD: *Unspecified part*: grazing, Australia (Parsons & Cuthbertson 1992), Brazil (Pio Corrêa 1926), New Zealand (Healy & Edgar 1980).
ENVIRONMENTAL USES: *Ornamentals*: (Grounds 1989; Huxley 1992); sometimes

sold in error as papyrus (Healy & Edgar 1980).

WEEDS: *Cultivation*: weed of beet fields, U.K. (H4); *Rice fields*: Australia (Parsons & Cuthbertson 1992; H68); *Waste places*: Australia (Parsons & Cuthbertson 1992; H70); *Irrigation ditches*: drains, channels, Australia and New Zealand (Healy & Edgar 1980; Parsons & Cuthbertson 1992); *Gardens*: (H69); *Pasture*: Australia (Parsons & Cuthbertson 1992).

Cyperus erythrorrhizos *Muhl.*, Descr. Gram.: 20 (1817).

N America; damp, sandy soils on margins of fresh water.

ENVIRONMENTAL USES: *Ornamentals*: (Huxley 1992).

Cyperus esculentus *L.*, Sp. Pl. 1: 45 (1753).

Tropical and warm temperate regions worldwide; cultivated in some regions; seasonally wet habitats, swamps, cultivated areas.

FOOD: *Tubers*: cultivated in Europe, Middle East, northern and southern Africa, Indo-China (Burkill 1935; Heywood 1993; Schermerhorn & Quimby 1960); sweet or nutty taste (De Vries 1991; Tredgold 1986); eaten raw (Anon. 1988; Peters 1992) or cooked (Täckholm & Drar 1950; Tredgold 1986); after roasting and grinding used as substitute for coffee and cocoa (Burkill 1935; Cauis & Banby 1935); sometimes made into caffeine-free 'chocolate' (Burkill 1985); used as substitute for almonds in confectionery (Anon. 1988; Burkill 1985); sold as 'tiger nuts', U.K. (H301 — Fig. 3E); used to make a sweet (Burkill 1985); used in preparation of a frozen or chilled drink known as 'horchata' or 'horchata de chufas' (Anon. 1988; Burkill 1985); boiled with wheat flour and sugar to make 'tiger nut milk' (Burkill 1985); used to prepare a thick dessert (Abbiw 1990); ground flour mixed with sorghum to make porridge (Anon. 1988); flour used in bread-making, ice-cream and sherberts (Anon. 1988); important food during drought or famine (Burkill 1985; Campbell 1986); oil from tubers edible (Burkill 1985); *Rhizomes*: used as vegetable (Dwivedi & Pandey 1992).

FOOD ADDITIVES: *Tubers*: source of potash for softening and flavouring green leafy vegetables (Tredgold 1986).

ANIMAL FOOD: *Unspecified part*: (Täckholm & Drar 1950); much grazed by cattle, Senegal (H182); *Tubers*: fed to cattle and pigs (Burkill 1935, 1985; Anon. 1988); *Leaves/culms/aerial parts*: (Burkill 1985).

BEE PLANTS: *Inflorescence*: bees reported to visit inflorescence (Burkill 1985).

MATERIALS: *Fibres*: culm used to make rough ropes, Libya (Burkill 1985), sleeping mats (Anon. 1988); culms and leaves could be used for paper-making (Anon. 1988; Burkill 1985); *Other materials/chemicals*: oil used for soap-making (Burkill 1935; Burkill 1985; Anon. 1988); a lubricating oil for watches etc., Egypt (Täckholm & Drar 1950).

SOCIAL USES: *Smoking materials and drugs*: rhizomes burnt to produce a scented smoke, N Nigeria (Burkill 1985); *Miscellanous social uses*: aphrodisiac, Madagascar (Cauis & Banby 1935), Sierra Leone (Burkill 1985), Egypt (Täckholm & Drar 1950), Ghana (Abbiw 1990).

MEDICINES: *Unspecified medicinal disorders*: used medicinally in Europe, W Africa, S Africa and Madagascar (Cauis & Banby 1935; Täckholm & Drar 1950); *Digestive system disorders*: tubers said to cure constipation, Ghana (Abbiw 1990); rhizome chewed for indigestion and bad breath (Cauis & Banby 1935, Burkill 1985, Tredgold

1986); rhizome a remedy for colic and hypochondriasis, Egypt (Cauis & Banby 1935); *Genito-urinary system disorders*: tubers used to help the start of menstruation in girls, S Africa (Burkill 1985; Cauis & Banby 1935); *Infections/ infestations*: tuber useful in treating eye problems, burning sensations and leprosy (Cauis & Banby 1935); *Metabolic system disorders*: tuber affects both heat production and heat regulation in the body (Cauis & Banby 1935); *Pain*: leaves used in headache and migraine, Guinea (Cauis & Banby 1935), Senegal (Burkill 1985); *Pregnancy/birth/puerperium disorders*: tuber used to increase milk supply in nursing mothers, Ivory Coast (Burkill 1985), Egypt (Täckholm & Drar 1950), Ghana (Abbiw 1990); oil an effective emollient for inflamed breasts of nursing mothers, Egypt (Täckholm & Drar 1950); *Sensory system*: tuber acting as a stimulant, Madagascar (Cauis & Banby 1935), China (Burkill 1985), Egypt (Täckholm & Drar 1950).

ENVIRONMENTAL USES: *Ornamentals*: (Huxley 1992).

WEEDS: *Unspecified weed*: (Schermerhorn & Quimby 1960); *Cultivation*: (Anon. 1988; Burkill 1985; Chivinge 1992; De Vries 1991; Fosberg 1988; Healy & Edgar 1980; H83, H92, H93, H183); *Rotation crops*: (Kühn 1982); *Perennial crops*: (Kühn 1982); *Waste places*: (Kühn 1982); *Gardens*: (Healy & Edgar 1980).

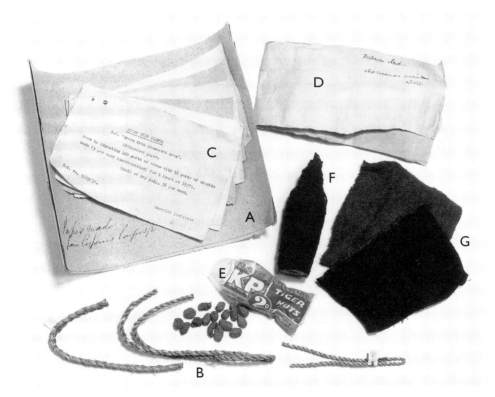

FIG. 3. **A** *Cyperus longus* paper, Channel Is. (EBMC 34320); **B** *Cyperus corymbosus* rope, China (EBMC 34239); **C** *Cyperus papyrus* paper, Sudan (EBMC 34230); **D** *Lepidosperma gladiatum* paper, Australia (EBMC 34235); **E** *Cyperus esculentus* 'tiger nuts', U.K. (EBMC 34335); **F** *Eriophorum vaginatum* yarn, U.K. (EBMC 34352); **G** *Eriophorum vaginatum* fabric, U.K. (EBMC 34354). All from the Economic Botany Museum collections, Kew; catalogue number in brackets.

Cyperus exaltatus *Retz.*, Observ. Bot. 5: 11 (1789).

Pantropical; wet or swampy places.

FOOD ADDITIVES: *Unspecified part*: burnt to produce a vegetable salt, Chad (Burkill 1985).

ANIMAL FOOD: *Unspecified part*: eaten by elephant, India (H38).

MATERIALS: *Fibres*: possible use in local paper-making (Burkill 1935); culms used for hut-building (Burkill 1985), thatching, Kenya (Burkill 1985), Tanzania (H239, H241); used for matting, India (Burkill 1935).

MEDICINES: *Infections/infestations*: rhizome used on infected skin swellings, Tanzania (Burkill 1985); rhizome used dressing scarifications over the spleen in chronic malaria, Tanzania (Burkill 1985); *Pregnancy/birth/puerperium disorders*: rhizome used to promote milk-flow in nursing mothers, Tanzania (Burkill 1985).

WEEDS: *Rice fields*: Tanzania (H241); *Irrigation ditches*: Ethiopia (H225), Kenya (H240).

Cyperus fastigiatus *Rottb.*, Descr. Icon. Rar. Pl.: 32, t. 7, f. 2 (1773).

S Africa; permanently wet and periodically inundated areas.

MEDICINES: *Unspecified medicinal disorders*: S Africa (Cauis & Banby 1935).

Cyperus fertilis *Boeck.*, Bot. Jahrb. Syst. 5: 90 (1884).

Tropical Africa; damp or wet places in forests.

ENVIRONMENTAL USES: *Ornamentals*: (Huxley 1992).

Cyperus fischerianus *A. Rich.*, Tent. Fl. Abyss. 2: 488 (1851).

E and NE tropical Africa; forest edges, open places in upland forest.

ANIMAL FOOD: *Unspecified part*: eaten by sheep and goats, Sudan (H227).

Cyperus flavus (*Vahl*) *Nees*, Linnaea 19: 698 (1847).

Mariscus flavus Vahl, Enum. Pl. 2: 374 (1806).

Warm-temperate to tropical America; open sandy soils, pastures, roadsides, waste places.

WEEDS: *Rotation crops*: (Kühn 1982); *Perennial crops*: (Kühn 1982); *Waste places*: (Kühn 1982).

Cyperus foliaceus *C. B. Clarke*, Kew Bull. Addit. Ser. 8: 5 (1908).

Tropical Africa; wet places.

WEEDS: *Rice fields*: Gambia (H205), Sierra Leone (H204).

Cyperus fulgens *C. B. Clarke*, Bull. Herb. Boiss. 4 App. 3: 30 (1896).

Tropical and subtropical subsaharan Africa; seasonally wet grasslands.

FOOD: *Tubers*: eaten raw or cooked (Arnold *et al.* 1985; Fox & Norwood Young 1982; Peters 1992); sweet taste, eaten raw, Zambia (H274); important in Kalahari, S Africa and Kaokoland, Namibia; tubers roasted in hot ash and boiled in milk, Botswana, S Africa (Van Eyk & Gericke 2000).

Cyperus fuscus *L.*, Sp. Pl. 1: 46 (1753).

Europe and N Africa to Indo-China, N America; wet places.

ANIMAL FOOD: *Unspecified part*: somewhat grazed, Turkey (H120).

MEDICINES: *Digestive system disorders*: used for suppressing flatulence (Kükenthal 1935 – 1936).

WEEDS: *Rice fields*: Egypt (H121); Iraq (H122).

Cyperus giganteus *Vahl*, Enum. Pl. 2: 364 (1805).
Central Americal, W Indies, tropical S America; marshes or standing water.

MATERIALS: *Fibres*: used for matting, the fibres similar to linen and rami, Brazil (Pio Corrêa 1926); used for making ordinary and high quality translucent paper, deserves an important place in the paper industry, Brazil (Pio Corrêa 1926).

ENVIRONMENTAL USES: *Pollution control*: water purifier, Brazil (Pio Corrêa 1926).

Cyperus gioli *Chiov.*, Ann. Bot. (Rome) 13: 375 (1915).
Somalia; irrigated arable land, wet places.

FOOD: *Rhizomes*: cooked and uncooked vegetable, herb, famine food, Somalia (Anon. 1994 – 2000).

Cyperus glomeratus *L.*, Cent. Pl. 2: 5 (1756).
Europe to N India and China; open, swampy areas.

MATERIALS: *Other materials/chemicals*: sweet-smelling rhizome used in perfume and cosmetics (Kükenthal 1935 – 1936).

Cyperus gracilis *R. Br.*, Prodr.: 213 (1810).
Australia, New Caledonia; sandy or loamy soils in light shade, often in open woodland.

WEEDS: *Gardens*: Australia (H155).

Cyperus grandibulbosus *C. B. Clarke* in Dyer, Fl. Trop. Afr. 8: 353 (1902).
E tropical Africa; seasonally wet areas, often on sandy soil.

WEEDS: *Unspecified weed*: Kenya (H256).

Cyperus grandis *C. B. Clarke* in T. Durand & Schinz, Consp. Fl. Afr. 5: 564 (1894); in Dyer, Fl. Trop. Afr. 8: 372 (1902).
Tropical East Africa, Madagascar; in standing or moving water.

MATERIALS: *Fibres*: thatching, Tanzania (H174).

Cyperus haspan *L.*, Sp. Pl. 1: 45 (1753).
Pantropical; wet and swampy places.

FOOD ADDITIVES: *Unspecified part*: burned to provide a vegetable salt, E Africa (Burkill 1985), Ghana (Abbiw 1990).

ANIMAL FOOD: *Unspecified part*: cattle, Tropical Africa, SE Asia (Burkill 1935; Burkill 1985).

MATERIALS: *Fibres*: mats and baskets, Togo (Kükenthal 1935 – 1936); used for mats and matting, E Africa (Greenway 1950); *Other materials/chemicals*: potash salts, Tropical Africa (Burkill 1935; Burkill 1985); pith used for lamp-wicks, Malaysia (Burkill 1935; Burkill 1985).

MEDICINES: *Unspecified medicinal disorders*: whole plant, particularly rhizome, used with other febrifuge plants by Wayãpi, French Guiana (Milliken 1997).

ENVIRONMENTAL USES: *Ornamentals*: (Huxley 1992).

WEEDS: *Unspecified weed*: (H59); *Rotation crops*: soya bean fields, Thailand (Radanachaless & Maxwell 1994); *Grassland*: (Kühn 1982); *Aquatic biotopes*: (Kühn 1982); *Rice fields*: (Kühn 1982; Simpson & Koyama 1998).

Cyperus hemisphaericus *Boeck.*, Flora 43: 436 (1859).
E and S tropical Africa; seasonally wet grasslands and savannahs.

FOOD: *Rhizomes*: cooked and uncooked vegetable and famine food (Anon. 1994 – 2000).

Cyperus imbricatus *Retz.*, Observ. Bot. 5: 12 (1789).
Pantropical; swampy areas.

ANIMAL FOOD: *Unspecified part*: grazing, Sudan (Burkill 1985).

MATERIALS: *Fibres*: culms used for mats and screens, Philippines (Burkill 1935); used to make string, Java (Kern 1974).

SOCIAL USES: *Miscellaneous social uses*: crushed rhizome used as an aphrodisiac by the Shipibo- Conibo, Peru (Tournon *et al.* 1986).

WEEDS: *Aquatic biotopes*: (Kühn 1982); *Rice fields*: (Kühn 1982; Simpson & Koyama 1998).

Cyperus immensus *C. B. Clarke*, J. Linn. Soc., Bot. 20: 294 (1883).
E and S tropical Africa; swampy areas, river banks.

MATERIALS: *Fibres*: used as a tying material and for thatch, E Africa (Greenway 1950).

Cyperus iria *L.*, Sp. Pl. 1: 45 (1753).
Pantropical; open wet places.

ANIMAL FOOD: *Unspecified part*: cattle and other livestock, Malay Peninsula, Hong Kong, India, Macau, Sudan (Burkill 1935; Burkill 1985; Mukhopadhyay & Ghosh 1992; Sudan Ministry 1980; H34).

MATERIALS: *Fibres*: matting, India (Mukhopadhyay & Ghosh 1992), Malaysia (Burkill 1985); *Other materials/chemicals*: insect juvenile hormone 'III', isolated from the plant, represents a valuable source for research purposes, Malaysia (Toong *et al.* 1988).

SOCIAL USES: '*Religious*' *uses*: (H53).

MEDICINES: *Infections/infestations*: drunk for fever, ground with *C. rotundus* tubers, India (Cauis & Banby 1935; H45); *Nervous system disorders*: taken as a stimulant and tonic, India (Millar & Morris 1988).

WEEDS: *Unspecified weed*: (Abbiw 1990; H50); *Cultivation*: sugar cane fields, Australia (H67), maize fields, India (Dangol 1992); *Rotation crops*: soya bean fields, Thailand (Radanachaless & Maxwell 1994); *Waste places*: (Kühn 1982); *Aquatic biotopes*: (4); *Rice fields*: (Burkill 1985; Kühn 1982; Simpson & Koyama 1998; H24, H47, H111, H167); *Gardens*: (Burkill 1935; Cauis & Banby 1935; H45).

Cyperus javanicus *Houtt.*, Nat. Hist. II, 13: Aanw. Pl. (1), t. 88, f. 1 (1782).

Tropical Africa and Madagascar, India and Sri Lanka to SE China and the Ryukyu Is., Malesia, Pacific Is. and W Australia; swampy places.

MATERIALS: *Fibres*: culm fibres used for coconut strainers, cordage and rope (Funk 1978); *Other materials/chemicals*: culms used to filter coconut milk and traditional medicinal preparations, French Polynesia, Marquesas Is. (H104).

MEDICINES: *Infections/infestations*: inflorescence pounded with coconut oil rubbed on the body as diaphoretic agent in cold, fever and malaria, Nicobar Is., India (Dagar & Dagar 1986, 1999).

WEEDS: *Cultivation*: (H103).

Cyperus jeminicus *Rottb.*, Descr. Icon. Rar. Pl.: 25, t. 8, f. 1 (1773).
Cyperus conglomeratus Rottb. var. *multiculmis* (Boeck.) Kük. in Engl., Pflanzenr. 4 (20), 101 Heft: 274 (1936).
Arabia, N Africa; sandy soils in open areas.

ANIMAL FOOD: *Unspecified part*: grazing, cattle, Senegal, Tunisia (Burkill 1985); grazing value fair, useful in first part of dry season, Sudan (H230); grazed by camels, Sudan (H231).

ENVIRONMENTAL USES: *Erosion control*: potential for sand stabilisation, Niger (H220).

Cyperus laevigatus *L.*, Mant. Pl. 2: 179 (1771).
Widespread in the tropics and subtropics; saline areas, hot springs, swamp margins and margins of water bodies.

FOOD ADDITIVES: *Unspecified part*: burnt for salt, Sudan (H232), Uganda (H265).

ANIMAL FOOD: *Unspecified part*: S Africa (H27), Australia (H72); *Leaves/culms/aerial parts*: culm grazed by cattle (Anon. 1994 – 2000), Hawaii (Jobe 1991); grazed by sheep and goats, Sudan (H233), grazed by cattle, sheep, donkeys, camels, Kenya (H266).

MATERIALS: *Fibres*: culms used for matting (Funk 1978), S Sandwich Is. (H79), cultivated in Hawaii to make mats, hats, cloaks, loincloths (Jobe 1991; P. van Dyke pers. comm. 1994); culms used for string, E Africa (Greenway 1950); *Unspecified medicinal disorders*: used by Masai, Kenya (H267).

WEEDS: *Aquatic biotopes*: (Kühn 1982).

Cyperus latifolius *Poir.* in Lam., Encycl. 7: 268 (1806).
Tropical Africa, Mascarenes; swampy areas, roadside ditches, stream-sides.

MATERIALS: *Fibres*: culms used for thatching, Tanzania (Burkill 1985); sleeping mats, S Africa (Van Eyk & Gericke 2000); craftwork resource, has been planted in paddy fields, S Africa (Heinsohn 1990).

SOCIAL USES: *'Religious uses'*: used for building totem/spirit houses, Kenya (H247), Tanzania (H117).

WEEDS: *Pasture*: Uganda (H246).

Cyperus laxus *Lam.*, Tabl. Encycl. 1: 146 (1791).
Cyperus diffusus Vahl, Enum. Pl. 2: 321 (1806).
India to SE China and Taiwan, eastwards to Malesia and Solomon Is; forests, river-banks, roadsides, usually in shade.

SOCIAL USES: *Miscellaneous social uses*: crushed rhizome used as an aphrodisiac by the Shipibo-Conibo, Peru (Tournon *et al.* 1986).

MEDICINES: *Unspecified medicinal disorders*: whole plant, particularly rhizome, used with other febrifuge plants by Wayãpi, French Guiana (Milliken 1997); *Infections/infestations*: rhizome used for lip disease known as 'singao, Philippines (Altschul 1973; Kern 1974).

var. **macrostachyus** (*Boeck.*) *Karth.* in Karth. *et al.*, Fl. Ind. Enum.-Monocot.: 46 (1989).
Cyperus diffusus Vahl var. *macrostachys* Boeck., Linnaea 35: 534 (1868).
Scattered throughout the range of var. *laxus*; forests, river-banks, roadsides, usually in shade.
WEEDS: *Gardens*: (H150, H151).

Cyperus ligularis *L.*, Amoen. Acad. 5: 391 (1760).
Tropical Africa, central and S America; sandy soils in open or light shade, creek margins.
ANIMAL FOOD: *Unspecified part*: forage, Brazil (Pio Corrêa 1926).
MATERIALS: *Other materials/chemicals*: culms used to make brushes to apply whitewash to houses, Ghana (Burkill 1985).

Cyperus longibracteatus (*Cherm.*) *Kük.*, Repert. Spec. Nov. Regni Veg. 26: 250 (1929).
Mariscus longibracteatus Cherm., Bull. Mus. Hist. Nat. (Paris) 25: 407 (1919).

var. **rubrotinctus** (*Cherm.*) *Kük.*, Repert. Spec. Nov. Regni Veg. 26: 250 (1929).
Mariscus rubrotinctus Cherm., Bull. Mus. Hist. Nat. (Paris) 25: 407 (1919).
Tropical Africa, Madagascar; river margins, damp to wet soils on margins of cultivation.
ANIMAL FOOD: *Unspecified part*: domestic livestock, Kenya (Burkill 1985); cattle grazing, Kenya (H185, H186, H188).
MATERIALS: *Fibres*: culms used for thatching, Kenya (H188); mat and basket weaving, Kenya (Burkill 1985; H186); culms used to make fish-nets, C African Republic (H223); *Other materials/chemicals*: perfumery (Anon. 1994 – 2000).

Cyperus longus *L.*, Sp. Pl. 1: 45 (1753).
Europe, Middle East to India, N Africa; wet places.
MATERIALS: *Fibres*: paper, Channel Is. (H293 — Fig. 2G, H297 — Fig. 3A), rope, Channel Is. (H296 — Fig. 2K); *Other materials/chemicals*: rhizomes used in perfumery (Heywood 1993; Kükenthal 1935 – 1936), Europe (Burkill 1985).
SOCIAL USES: *Miscellaneous social uses*: culm bases used as an aphrodisiac (Kükenthal 1935 – 1936).
VERTEBRATE POISONS: *Mammals*: sap poisonous, burns skin, Zimbabwe (Cauis & Banby 1935).
MEDICINES: *Unspecified medicinal disorders*: used medicinally, Europe and S Africa (Cauis & Banby 1935); rhizomes, veterinary (Anon. 1994 – 2000); rhizome mucilage

aromatic and bitter with tonic properties, U.K. (Pratt 1900); *Digestive system disorders*: rhizome used in enema from for children with stomach problems, S Africa (Cauis & Banby 1935); culm bases used to enhance digestion (Kükenthal 1935 – 1936); rhizome mucilage with stomachic properties, U.K. (Pratt 1900); *Genito-urinary system disorders*: rhizome decoction diuretic (Täckholm & Drar 1950); *Infections/infestations*: rhizome chewed or powdered rhizome blown into nose and ears for colds, S Africa (Cauis & Banby 1935; Kükenthal 1935 – 1936); *Metabolic system disorders*: (Kükenthal 1935 – 1936); *Muscular-skeletal system disorders*: rhizome decoction used for rheumatism, Egypt (Täckholm & Drar 1950); *Pregnancy/ birth/puerperium disorders*: culm bases used as a febrifuge for easing labour (Kükenthal 1935 – 1936).

ENVIRONMENTAL USES: *Ornamentals*: (Grounds 1989; Huxley 1992).

WEEDS: *Aquatic biotopes*: (Kühn 1982); *Rice fields*: (Kühn 1982).

var. pallidior *Kük.* in Engl., Pflanzenr. 4 (20), 101 Heft: 100 (1935).
S Europe, Middle East, SW Asia to India; wet places.

WEEDS: *Rice fields*: Iran (H117), Iraq (H118).

var. pallidus *Boeck.*, Linnaea 36: 280 (1870).
Tropical Africa: periodically flooded areas in grasslands.

ANIMAL FOOD: *Unspecified part*: grazing, E Africa (Burkill 1985).

MATERIALS: *Other materials/chemicals*: poison-bait for locust-control, Somalia (Burkill 1985).

var. tenuiflorus (*Rottb.*) *Boeck.*, Linnaea 36: 281 (1870).
C. tenuiflorus Rottb., Descr. Icon. Rar. Pl.: 30, t. 14, f. 1 (1773).
Subsaharan Africa; wet places, periodically flooded areas.

WEEDS: *Irrigation ditches*: Kenya (Burkill 1985).

Cyperus luzulae (*L.*) *Retz.*, Observ. Bot. 4: 11 (1786).
Scirpus luzulae L., Syst. Nat. ed. 10, 2: 868 (1759).
Tropical and subtropical America; stream sides and other damp places.

ANIMAL FOOD: *Unspecified part*: inferior quality forage, Brazil (Pio Corrêa 1926).

SOCIAL USES: *Miscellaneous social uses*: crushed rhizome used as an aphrodisiac by the Shipibo-Conibo, Peru (Tournon *et al.* 1986).

MEDICINES: *Skin/subcutaneous cellular tissue disorders:* used by the Shipibo-Conibo to maintain and encourage hair growth, mixed with *Genipa americana* L. (*Rubiaceae*) and rubbed on to the hair, Peru (Tournon *et al.* 1986).

WEEDS: *Cultivation*: Brazil (Pio Corrêa 1926).

Cyperus maculatus *Boeck.* in Peters, Reise Mossamb. Bot. 2: 539 (1864).
Africa, Madagascar; sandy areas near water.

FOOD: *Culms*: used as cooked and uncooked vegetable, famine food (Anon. 1994 – 2000); *Rhizomes*: edible (Burkill 1985).

ANIMAL FOOD: *Leaves/culms/aerial parts*: foliage, grazing, Sudan and Senegal (Burkill 1985).

MATERIALS: *Other materials/chemicals*: rhizomes sold in fragrant sachets and perfume, burnt in fires to create a pleasant odour (Burkill 1985), Nigeria (H116).

MEDICINES: *Infections/infestations*: infusion of leaves and rhizomes put into a preparation for treating 'garli' cattle disease, Ghana (Burkill 1985).

Cyperus malaccensis *Lam.*, Tabl. Encycl. 1: 146 (1791).
Middle East to Indo-China and S China, Malesia to N Australia and Polynesia; marshy places near brackish or salt water.

ANIMAL FOOD: *Unspecified part*: cattle and water buffalo, Iraq (H98).

MATERIALS: *Fibres*: culms used for matting (Heywood 1993; Lewington 1990), Malay Peninsula, Philippines, Indonesia (Burkill 1935), China (H29), Iraq (H97), Papua New Guinea (Kükenthal 1935 – 1936; H55); culms used for cordage and ropes (Burkill 1935), Philippines (H52), Vietnam (Nguyen 1993); culms used for hats and baskets (Kern 1974); pieces of culm plaited in ropes which are used to attract fish fry, Java (Kern 1974); *Other materials/chemicals*: rhizome used as a fumigant, Iraq (H97).

MEDICINES: *Genito-urinary system disorders*: rhizomes considered to be diuretic, Vietnam (Nguyen 1993).

WEEDS: *Cultivation*: Iraq (H99); *Aquatic biotopes*: (Kühn 1982).

Cyperus mapanioides *C. B. Clarke* in T. Durand & Schinz, Consp. Fl. Afr. 5: 568 (1894); in Dyer, Fl. Trop. Afr. 8: 340 (1901).
Tropical Africa; forest and cultivated areas.

WEEDS: *Cultivation*: Nigeria (H217); *Rotation crops*: Congo (H219); *Gardens*: Congo (H218).

Cyperus maranguensis *K. Schum.* in Engl., Pflanzenwelt Ost.-Afr. C.: 120 (1895).
Tropical E Africa; cultivated areas, roadsides, grassland, forests.

MATERIALS: *Fibres*: basketry, E Africa (Greenway 1950).

WEEDS: *Cultivation*: (Haines & Lye 1983).

Cyperus margaritaceus *Vahl*, Enum. Pl. 2: 307 (1806).
Subsaharan Africa; open woodland or grassland, often on sandy soil.

ANIMAL FOOD: *Unspecified part*: young plant, Malawi (Burkill 1985).

MEDICINES: *Digestive system disorders*: rhizome laxative and purgative, Congo (Kinshasa) (H221).

ENVIRONMENTAL USES: *Ornamentals*: Nigeria (H213).

Cyperus marginatus *Thunb.*, Prodr. Fl. Cap. 1: 18 (1794).
Namibia, S Africa; stream banks, pool margins.

MATERIALS: *Fibres*: thatching, Namibia (Van Eyk & Gericke 2000).

Cyperus michelianus (*L.*) *Link*, Hort. Bot. Berol. 1: 303 (1827).
Scirpus michelianus L., Sp. Pl. 1: 52 (1753).

subsp. **pygmaeus** (*Rottb.*) *Asch. & Graebn.*, Syn. Mitteleur. Fl. 2, 2: 273 (1904).
Cyperus pygmaeus Rottb., Descr. Icon. Rar. Pl.: 20, t. 14, f. 4 & 5 (1773).

Mediterranean and E Africa through W Asia and India to SE Asia, Malesia and Australia; open wet places.

ANIMAL FOOD: *Unspecified part*: very useful fodder plant, Iraq (H133).

WEEDS: *Cultivation*: (Simpson & Koyama 1998); *Gardens*: Iraq (H134, H135).

Cyperus multifolius *Kunth*, Enum. Pl. 2: 91 (1837).
Western tropical S America: damp places.

MATERIALS: *Other materials/chemicals*: rhizomes used in perfumery, Colombia (H312).

Cyperus natalensis *Hochst.*, Flora 28: 755 (1845).
Southern Africa; open, sandy coastal areas.

MATERIALS: *Fibres*: sleeping mats, collecting baskets, winnowing baskets, grinding mats, rolled twine, cutting boards, S Africa (Van Eyk & Gericke 2000).

Cyperus nutans *Vahl*, Enum. Pl. 2: 363 (1806).
India, Sri Lanka, Indo-China, S China and Malesia; swamps, wet open places, and wet forest.

MATERIALS: *Fibres*: matting, Malaysia (Burkill 1935).

WEEDS: *Rotation crops*: soya bean fields, Thailand (Radanachaless & Maxwell 1994); *Rice fields*: (Simpson & Koyama 1998).

subsp. **eleusinoides** (*Kunth*) *T. Koyama*, Gard. Bull. Singapore 30: 136 (1977).
Cyperus eleusinoides Kunth, Enum. Pl. 2: 39 (1835).
Tropical Africa, India to Malesia and N Australia, also N to the Ryukyu Is; swamps, wet open places, and wet forest.

MATERIALS: *Fibres*: culms used for making rope, India (H140).

Cyperus obtusiflorus *Vahl*, Enum. Pl. 2: 308 (1806).
C, E and S tropical Africa; dry grasslands, savannahs, rocky slopes,

ANIMAL FOOD: *Unspecified part*: grazed by cattle, Kenya (H261).

WEEDS: *Gardens*: Tanzania (H262).

Cyperus odoratus *L.*, Sp. Pl. 1: 46 (1753).
Cyperus ferax Rich., Actes Soc. Hist. Nat. Paris 1: 106 (1792).
Mariscus ferax (Rich.) C. B. Clarke in Hook. f., Fl. Brit. India 6: 624 (1893).
Torulinium odoratum (L.) S. S. Hooper, Kew Bull. 26: 579 (1972).
Pantropical; open wet, marshy or grassy places, waysides and wet sandy places near coasts.

MATERIALS: *Fibres*: matting, Sulawesi (Kern 1974), E Africa (Greenway 1950).

SOCIAL USES: *Miscellaneous social uses*: crushed rhizome used as an aphrodisiac by the Shipibo-Conibo, Peru (Tournon *et al.* 1986).

MEDICINES: *Unspecified medicinal disorders:* antispasmodic and stomachic properties, Brazil (Pio Corrêa 1926).

WEEDS: *Cultivation*: (Svenson 1943; H107); *Waste places*: (Kühn 1982); *Aquatic biotopes*: (Kühn 1982); *Rice fields*: (Kühn 1982; Simpson & Koyama 1998).

Cyperus owanii *Boeck.*, Flora 61: 29 (1878).
Southern Africa: wet shady places in forest and on forest margins.
ENVIRONMENTAL USES: *Ornamentals*: (Huxley 1992).

Cyperus pangorei *Rottb.*, Descr. Icon. Rar. Pl.: 31, t. 7, f. 3 (1773).
India, Myanmar, Indonesia; open wet places, margins of water bodies, marshes.
MATERIALS: *Fibres*: matting, India, Myanmar (Amalraj 1991; Burkill 1935; Hooker 1857; Lewington 1990; H317); mats and baskets, Mauritius, Sulawesi (Kükenthal 1935 – 1936).

Cyperus papyrus *L.*, Sp. Pl.: 47 (1753).
Tropical Africa; widely cultivated; swamps, margins of water bodies.
FOOD: *Unspecified part*: (Burkill 1985); *Rhizomes*: eaten raw or cooked (Burkill 1935; Burkill 1985; Kükenthal 1935 – 1936; Millar & Morris 1988; Täckholm & Drar 1950); *Culms*: eaten raw or cooked (Burkill 1985; Kükenthal 1935 – 1936; Täckholm & Drar 1950); thick pith chewed as sugar cane, Botswana (Tredgold 1986), Zambia (S.A. Renvoize pers. comm. 1996).
ANIMAL FOOD: *Unspecified part*: grazed by cattle, especially in dry season (Muthuri & Kinyamario 1989); *Inflorescences/infructescences/nutlets*: eaten by livestock, Oman (Millar & Morris 1988).
MATERIALS: *Fibres*: culms used for paper making (Altschul 1973; Anon. 1994 – 2000; Burkill 1935; Burkill 1985; Huxley 1992; Kükenthal 1935 – 1936; Lewington 1990; Schermerhorn & Quimby 1960; Täckholm & Drar 1950, Tredgold 1986), Sudan (H299 — Fig. 3C); culms used for making fibre-board, Uganda (H310); culms used for cordage and ropes (Burkill 1935; Kükenthal 1935 – 1936; Täckholm & Drar 1950); culms used in boatmaking, Africa (Burkill 1985; Kükenthal 1935 – 1936; Lewington 1990; Täckholm & Drar 1950; H80, H81); culms used for matting (Täckholm & Drar 1950), mattresses and cushions (Burkill 1985), roofing and flooring, Oman (Millar & Morris 1988); culms used to make sleeping mats, Botswana (Van Eyk & Gericke 2000);
FUELS: *Miscellaneous fuels*: woody rhizome, Egypt (Burkill 1985; Kükenthal 1935 – 1936), whole plant, Uganda (Jones 1983).
SOCIAL USES: *'Religious' uses*: chewed dried rhizomes used to ward off 'evil spirits', Gabon (Burkill 1985).
MEDICINES: *Digestive system disorders*: ash used to prevent spread of malignant ulcers, Egypt (Täckholm & Drar 1950); ash from burnt inflorescence used to treat rectal prolapse, Uganda (Adjanohoun *et al.* 1993); *Genito-urinary system disorders*: rhizome-decoction with sap of *Maytenus senegalensis* (*Celastraceae*) used to treat female sterility, Tanzania (Burkill 1985); ash from burnt inflorescence used to treat vaginal prolapse, Uganda (Adjanohoun *et al.* 1993); *Injuries*: ash used to heal wounds, Egypt (Täckholm & Drar 1950); *Sensory system*: ash used to treat certain eye diseases, Egypt (Täckholm & Drar 1950).
ENVIRONMENTAL USES: *Ornamentals*: (Grounds 1989; Huxley 1992; Kükenthal 1935 – 1936; Parsons & Cuthbertson 1992; Schermerhorn & Quimby 1960).
WEEDS: *Unspecified weed*: tropical Africa (Burkill 1985); *Aquatic biotopes*: tropical Africa (Abbiw 1990).

Cyperus pilosus *Vahl*, Enum. Pl. 2: 354 (1806).
Old World tropics and subtropics; wet grasslands, swamps and rice fields.
    ANIMAL FOOD: *Unspecified part*: cattle fodder (Burkill 1935).
    WEEDS: *Unspecified weed*: (H75); *Rice fields*: (Burkill 1935; Simpson & Koyama 1998).

Cyperus platyphyllus *Roem. & Schult.*, Syst. Veg. 2: 876 (1817).
India; marshy places.
    MEDICINES: *Ill defined symptoms*: tuber said to have tonic and stimulant properties (Cauis & Banby 1935).

Cyperus procerus *Rottb.*, Descr. Icon. Rar. Pl.: 29, t. 5, f. 3 (1773).
Madagascar, India, Sri Lanka, Indo-China, SE China, Taiwan, Malesia and N Australia; open, swamps and wet places, sometimes in brackish marshes.
    MATERIALS: *Fibres*: culms used for making string, Malesia (Burkill 1935; Burkill 1985; Kern 1974); culms for making mats, India (H39, H333).
    WEEDS: *Rice fields*: (Burkill 1985).

var. **lasiorrhachis** *C. B. Clarke* in Hook. f, Fl. Brit. India 6: 610 (1893).
India; open, swamps and wet places, sometimes in brackish marshes.
    MATERIALS: *Fibres*: mats, India (Burkill 1985).

Cyperus prolifer *Lam.*, Tabl. Encycl. 1: 147 (1791).
E and S Africa, Madagascar; swamp and stream margins, seasonally flooded areas.
    MATERIALS: *Fibres*: baskets, Madagascar (H3).
    MEDICINES: *Unspecified medicinal disorders*: Madagascar (Cauis & Banby 1935).
    ENVIRONMENTAL USES: *Ornamentals*: (Huxley 1992).

    This species is often named in cultivation as *Cyperus papyrus* 'nanus' or *C. haspan* (Simpson 1994).

Cyperus prolixus *Kunth* in Humb., Bonpl. & Kunth, Nov. Gen. Sp. 1: 206 (1815).
C America, tropical and subtropical S America; sometimes cultivated; swamps, stream banks, open wet places in forest.
    MATERIALS: *Fibres*: used for weaving bottle covers, chairs seats and in other similar objects, Brazil (Pio Corrêa 1926); used to make very fine paper, Brazil (Pio Corrêa 1926).
    SOCIAL USES: *Smoking materials and drugs*: leaves and rhizomes mixed with tobacco, reportedly hallucinogenic when smoked, Peru (Plowman *et al.* 1990); *Antifertility agents*: rhizome a contraceptive and sterilant, Brazil (Plowman *et al.* 1990).
    MEDICINES: *Unspecified medicinal disorders*: important medicinal plant in the culture of Shuar and Achuar, Ecuador (Evans 1991); crushed rhizome applied to traumatised parts of body, Shipibo-Conibo, Peru (Tournon *et al.* 1986); *Digestive system disorders*: rhizome used to treat eating disorders, Ecuador (Plowman *et al.* 1990); rhizome used to treat diarrhoea, Venezuela (Plowman *et al.* 1990); plant parasitised with the fungus *Cintractia limitata* (*Ustilaginaceae*) is crushed in water and

taken in cases of diarrhoea, Shipibo-Conibo, Peru (Tournon *et al.* 1986); *Ill-defined symptoms*: to calm quick-tempered people, the rhizome is rubbed over the body, Shipibo-Conibo, Peru (Tournon *et al.* 1986); *Nutritional disorders*: rhizome used in preparation to help babies increase weight, Colombia (Plowman *et al.* 1990), *Pregnancy/birth/puerperium disorders*: purgative given to both wife and husband following the birth of a child, Ecuador (Plowman *et al.* 1990), culms and leaves used for hastening birth, Ecuador (Plowman *et al.* 1990).

Cyperus pseudosomaliensis *Kük.* in Engl., Pflanzenr. 4 (20), 101 Heft: 539 (1936).
Mariscus somaliensis C. B. Clarke, Kew Bull. 1895: 229 (1895); in Dyer, Fl. Trop. Afr. 8: 383 (1902).
Somalia; open grassy areas, cultivation.
    ANIMAL FOOD: *Unspecified part*: important grazing by all livestock, Somalia (H237).
    WEEDS: *Gardens*: Somalia (H238).

Cyperus pulcherrimus *Willd.* ex Kunth, Enum. Pl. 2: 35 (1837).
India, Sri Lanka, Indo-China, Malesia; open or shaded, wet or swampy localities such as wet forest floors, open pool margins and ditches, rice fields.
    WEEDS: *Rice fields*: (Simpson & Koyama 1998).

Cyperus pustulatus *Vahl*, Enum. Pl. 2: 341 (1806).
Tropical Africa; marshy areas, often near open water.
    WEEDS: *Aquatic biotopes*: Ghana (H194); *Rice fields*: Gambia (H190, H191), Togo (H193), Nigeria (H192).

Cyperus radians *Nees & Mey. ex Kunth*, Enum. Pl. 2: 95 (1837).
India, S and E China, W Malesia; coastal areas, particularly sea-shores and sand-dunes.
    WEEDS: *Waste places*: (Kühn 1982); *Aquatic biotopes*: (Kühn 1982).

Cyperus reduncus *Hochst. ex Boeck.*, Linnaea 35: 580 (1868).
Tropical Africa; seasonally wet areas.
    WEEDS: *Cultivation*: (H199); *Disturbed land*: Nigeria (H200); *Rice fields*: Gambia (H195), Togo (H196), Nigeria (H197, H198), Zambia (Burkill 1985).

Cyperus remotispicatus *S. S. Hooper*, Kew Bull. 26: 577 (1972).
W tropical Africa; damp places on disturbed ground
    WEEDS: *Rice fields*: Nigeria (H216).

Cyperus renschii *Boeck.*, Flora 65: 11 (1882).
Tropical Africa; swampy areas and stream-sides in forest.
    MEDICINES: *Circulatory system disorders*: sap from rhizome and leaf used to treat cardiac oedema, Tanzania (Burkill 1985).

var. **scabridus** *Lye*, Nordic J. Bot. 3 (2): 229 (1983).
Uganda, Tanzania, Congo (Kinshasa); cleared forest glades.

MEDICINES: *Digestive system disorders*: Infusion of rhizome used for stomach ache in children (H259).

Cyperus rigidifolius *Steud.*, Flora 25: 593 (1842).
E and S Africa; grasslands, roadsides and cultivated areas.
ANIMAL FOOD: *Unspecified part*: grazed by all domestic livestock, Kenya (H250).
MEDICINES: *Respiratory system disorders*: chest complaints and colds, Kenya (H253).
WEEDS: *Cultivation*: S Africa (H119); *Rotation crops*: Kenya (H251); *Gardens*: Uganda (H249); *Pasture*: Kenya (H252).

Cyperus rotundus *L.*, Sp. Pl. 1: 45 (1753).
Pantropical; open or slightly shaded areas; often in areas of cultivation.
FOOD: *Rhizomes*: edible, Ghana (Abbiw 1990), India (Wickens *et al.* 1989), Thailand (A. Thammathaworn & S. Thammathaworn pers. comm. 1998), Malay Peninsula (Burkill 1935), Australia (Parsons & Cuthbertson 1992); famine food (Burkill 1985); *Culms*: cooked vegetable (Anon. 1994 – 2000); culm bases edible (Kern 1974, Anon. 1994 – 2000, Peters 1992).
ANIMAL FOOD: *Unspecified part*: (Parsons & Cuthbertson 1992; H15); grazed by cattle (Burkill 1935), Oman (H15); grazed by buffalo, cows, asses, sheep and goats, India (Kulhari & Joshi 1992); fodder, Ghana (Abbiw 1990); *Rhizomes*: culm bases eaten by pigs, Ghana (Abbiw 1990); *Leaves/culms/aerial parts*: leaves (Kern 1974), Ghana (Abbiw 1990); young foliage eaten by grazing animals, particularly when food limited (Parsons & Cuthbertson 1992).
MATERIALS: *Other materials/chemicals*: culm bases used as bait for catching rats, Tanzania (Burkill 1985); culm bases used for incense and perfumery, W Africa (Burkill 1985), India (Parsons & Cuthbertson 1992), China (Chin & Keng 1992); culm bases used for traditional face powder, Sulawesi (H48); culm bases used as insect repellent, Middle East (Millar & Morris 1988), China (Chin & Keng 1992), Ghana (Abbiw 1990).
NON-VERTEBRATE POISONS: *Mollusca*: rhizome used for molluscicide (Anon. 1994 – 2000).
MEDICINES: *Unspecified medicinal disorders*: Europe, India, China, Indo-China, Malaysia and Philippines (Cauis & Banby 1935; Kern 1974; Parsons & Cuthbertson 1992); used as a tonic, India (Nguyen 1993; Shanmugasundaram *et al.* 1991); diaphoretic, diuretic, hypotensive and inflammatory due to presence of cyperone, Thailand (Saralamp *et al.* 1996); *Circulatory system disorders*: cardiac tonic, Thailand (Saralamp *et al.* 1996); *Digestive system disorders*: widespread use for treating stomach and bowel disorders including diarrhoea, indigestion, nausea, dysentry (Burkill 1935; Cauis & Banby 1935; Bhattari 1993; Chin & Keng 1992; Deokule & Magdum 1992; Jain 1992; Kapur *et al.* 1992a; Kükenthal 1935 – 1936; Mukhopadhyay & Ghosh 1992; Nguyen 1993; Shanmugasundaram *et al.* 1991; Täckholm & Drar 1950; Tiwari *et al.* 1992; Vedavathy 1991); culm bases used for liver disorders, China (Cauis & Banby 1935), Cambodia (Burkill 1935); *Endocrine system disorders*: culm bases used in herbal treatment for diabetes, India (Reddy *et al.* 1991); *Genito-urinary system disorders*: culm bases used as a diuretic (Burkill 1935, Parsons & Cuthbertson 1992),

Middle East (Millar & Morris 1988), India (Shanmugasundaram *et al.* 1991; Vedavathy 1991), Cambodia (Cauis & Banby 1935), Indo-China (Burkill 1935), Java (Burkill 1935; Nguyen 1993); rhizomes and culm bases used to regulate menstruation and treat menstrual disorders, Middle East, (Millar & Morris 1988), Ghana (Abbiw 1990), China (J. C. Shaw, C. S. Tong & L. Wong pers. comm. 1995; H33); *Infections/ infestations*: widespread use of culm bases in reducing fever (Burkill 1985; Cauis & Banby 1935; Kükenthal 1935 – 1936; Millar & Morris 1988; Mohsin *et al.* 1989; Parsons & Cuthbertson 1992; Reddy *et al.* 1991; Shanmugasundaram *et al.* 1991; Täckholm & Drar 1950, Vedavathy 1991); culm bases used against parasitic worms (Täckholm & Drar 1950), Middle East (Millar & Morris 1988), Philippines (Cauis & Banby 1935), India (Pal 1992a; Parsons & Cuthbertson 1992; Vedavathy 1991), China (Kükenthal 1935 – 1936); stolon used to treat whooping cough, Ghana (Abbiw 1990); culm bases used to treat malaria, Tanzania (Burkill 1985), Cambodia (Burkill 1935), inhibits growth of *Plasmodium falciparium* in vitro, Thailand (Saralamp *et al.* 1996); paste made from culm bases used to remove lice, India (Deokule & Magdum 1992); *Inflammation*: stolon decoction used in ophthalmia (Cauis & Banby 1935); anti-swelling, China (H33); *Mental disorders*: common ingredient in herbal remedies for insanity, hysteria, insomnia and anxiety, Nigeria (Adesina 1990); culm base used to improve memory, India (Shanmugasundaram *et al.* 1991); *Nervous system disorders*: culm bases have stimulant properties (Burkill 1985), India (Nguyen 1993; R. C. Srivastava pers. comm 1992), Philippines (Cauis & Banby 1935); stolon and culm bases used in herbal treatment for epilepsy (Cauis & Banby 1935), India (Reddy *et al.* 1991; Shanmugasundaram *et al.* 1991); *Nutritional disorders*: culm base improves taste, acts as a nourisher and used in anorexia, India (Shanmugasundaram *et al.* 1991); *Pain*: culm base used in general pain relief (Burkill 1935; Cauis & Banby 1935), China, (Nguyen 1993), Cambodia (Nguyen 1993); stolon and culm base used specifically to treat headache, Cambodia (Burkill 1935), China (Chin & Keng 1992), India (Chin & Keng 1992; Reddy *et al.* 1991); *Poisonings*: culm base used in treatment of scorpion stings and snake bite (Cauis & Banby 1935), India (Millar & Morris 1988; Shukla *et al.* 1992); *Pregnancy/birth/puerperium disorders*: rhizomes and culm bases beneficial in assisting labour (Burkill 1985; Kükenthal 1935 – 1936), Indo-China (Burkill 1935), Cambodia (Nguyen 1993); culm bases used as a poultice to encourage flow of breast-milk, India (Millar & Morris 1988; Shanmugasundaram *et al.* 1991); *Respiratory system disorders*: culm bases used to treat coughs, colds and bronchial asthma, China (Chin & Keng 1992; Parsons & Cuthbertson 1992), India (Parsons & Cuthbertson 1992; Shanmugasundaram *et al.* 1991), N Ghana (Burkill 1985); *Sensory system*: nutlets heated in oil and used to clear ear wax, Oman (Millar & Morris 1988); *Skin/subcutaneous cellular tissue disorders*: stolon used in mixtures for poulticing sores and ulcers, India and Indonesia (Burkill 1935).

ENVIRONMENTAL USES: *Erosion control*: used to bind soil as protection against wind erosion (Parsons & Cuthbertson 1992).

WEEDS: *Unspecified weed*: (Al-Zoghet 1989; Chin & Keng 1992; Huxley 1992; Parsons & Cuthbertson 1992; H6, H31, H32, H63, H76, H77); *Cultivation*: (Abbiw 1990; Burkill 1935; Burkill 1985; Chivinge 1992; Dangol 1992; Nguyen 1993; Simpson 1992a; H5, H8, H10, H16, H17, 861, H20, H21, H23, H28, H30); described as world's worst weed, troublesome in 52 crops in 92 countries (Parsons & Cuthbertson 1992); *Rotation crops*: soya bean fields, Thailand (Radanachaless & Maxwell 1994); *Perennial crops*: (Kühn 1982); *Grassland*: (Kühn 1982); *Waste places*: (Burkill 1935; Burkill 1985; Kühn 1982; H82); *Rice fields*: (H14, H22); *Irrigation ditches*: (Parsons & Cuthbertson 1992; H7); *Gardens*: (Burkill 1935; Parsons & Cuthbertson 1992; Simpson 1992a; H9, H11, H12, H13, H18, H19, H30, H40, H41, H56, H61, H62, H64, H83).

subsp. **merkeri** (*C. B. Clarke*) *Kük.* in Engl., Pflanzenr. 4 (20), 101 Heft: 115 (1935). *Cyperus merkeri* C. B. Clarke, Kew Bull. Addit. Ser. 8: 12 (1908). Tropical E and NE Africa; similar to the typical subspecies.
   ANIMAL FOOD: *Unspecified part*: grazed by cattle, Kenya (H254); grazed by donkeys and sheep, Kenya (H255).
   WEEDS: *Gardens*: Burundi (H206).

subsp. **retzii** (*Nees*) *Kük.* in Engl., Pflanzenr. 4 (20), 101 Heft: 114 (1935). *Cyperus retzii* Nees in Wight, Contr. Bot. India: 82 (1834). Africa, India, N Australia; as above.
   MEDICINES: *Infections/infestations*: culm used to treat gonorrhoea (Anon. 1994 – 2000).
   WEEDS: *Cultivation*: Australia (H152, H153).

subsp. **tuberosus** (*Rottb.*) *Kük.* in Engl., Pflanzenr. 4 (20), 101 Heft: 113 (1935). *C. tuberosus* Rottb., Descr. Icon. Rar. Pl.: 28, t. 7, f. 1 (1773). Old World tropics and subtropics; damp grasslands and waste places.
   FOOD: *Culms*: edible culm bases (Burkill 1935).
   ANIMAL FOOD: *Unspecified part*: cattle, Senegal (Burkill 1985), all livestock, Mauritania (Burkill 1985).
   WEEDS: *Unspecified weed*: (Abbiw 1990); Australia (H65); *Cultivation*: E Africa (Burkill 1985).
   NOTE: Subsp. *tuberosus* is very close to the typical subspecies. We have identified subsp. *tuberosus* where specifically mentioned in a reference but it is likely that some records of the typical subspecies could be referable to subsp. *tuberosus*.

Cyperus rubicundus *Vahl*, Enum. Pl. 2: 308 (1806). Tropical Africa; seasonally wet habitats, often on shallow soils.
   ANIMAL FOOD: *Unspecified part*: grazed by cattle, Kenya (H263).
   MEDICINES: *Digestive system disorders*: infusion of rhizome used to relieve stomach pain (H264).

Cyperus serotinus *Rottb.*, Descr. Pl. Rar.: 18 (1772); Descr. Icon. Rar. Pl.: 31 (1773). Europe, through central Asia to China, Taiwan and N Indo-China; wet marshy places.
   WEEDS: *Aquatic biotopes*: (Kühn 1982); *Rice fields*: (Kühn 1982).

var. **inundatus** (*Roxb.*) *Kük.* in Engl., Pflanzenr. 4 (20), 101 Heft: 318 (1936). *Cyperus inundatus* Roxb., Hort. Bengal.: 6 (1814) & Fl. Ind. 1: 201 (1832). India, China, Taiwan; wet places.

MEDICINES: *Ill defined symptoms*: (Cauis & Banby 1935).

Cyperus sexangularis *Nees*, Linnaea 9: 284 (1835).
Southern Africa; margins of rivers, streams and other wet places.
MATERIALS: *Fibres*: sleeping mats, S Africa (Van Eyk & Gericke 2000); craftwork resource, has been planted in paddy fields, S Africa (Heinsohn 1990).
MEDICINES: *Unspecified medicinal disorders*: S Africa (Cauis & Banby 1935).
ENVIRONMENTAL USES: *Ornamentals*: (Gordon-Gray 1995).

Cyperus soyauxii *Boeck.*, Bot. Jahrb. Syst. 5: 501 (1884).
Mariscus soyauxii (Boeck.) C. B. Clarke in T. Durand & Schinz, Consp. Fl. Afr. 5: 593 nom. nud; in Dyer, Fl. Trop. Afr. 8: 393 (1902).
W tropical Africa; waste places, forest margins, cultivated land.
MATERIALS: *Other materials/chemicals*: rhizome aromatic, Ghana (H211).
WEEDS: *Cultivation*: Ghana (H212).

Cyperus sphacelatus *Rottb.*, Descr. Pl. Rar.: 21 (1772); Descr. Icon. Rar. Pl.: 26 (1773).
Tropical America and tropical Africa; introduced elsewhere; open grasslands, river banks, cultivated areas.
ANIMAL FOOD: *Unspecified part*: sheep and goats, Ghana (Abbiw 1990; Burkill 1985).
WEEDS: *Cultivation*: Sierra Leone (H85, H86); *Rotation crops*: (Kühn 1982); *Grassland*: Nigeria: (H90); *Waste places*: (Kühn 1982); *Disturbed land*: Nigeria (H88); *Aquatic biotopes*: (Kühn 1982); *Rice fields*: Gambia (H84); *Gardens*: Nigeria (H89); *Forests*: Ghana (H87).

Cyperus squarrosus *L.*, Cent. Pl. 2: 6 (1756).
Tropical to warm temperate regions worldwide; seasonally wet areas.
ANIMAL FOOD: *Unspecified part*: cattle, Senegal (Burkill 1985).
MATERIALS: *Other materials/chemicals*: sweet-smelling rhizome used in perfumery and cosmetics (Kükenthal 1935 – 1936).
MEDICINES: *Genito-urinary system disorders*: entire plant used to treat impotence, Uganda (Adjanohoun *et al.* 1993).
WEEDS: *Cultivation*: (Burkill 1985), *Rotation crops*: (Kühn 1982); maize fields, India (Dangol 1992); *Perennial crops*: (Kühn 1982); *Grassland*: (Kühn 1982); *Waste places*: (Kühn 1982); *Aquatic biotopes*: (Kühn 1982).

Cyperus stoloniferus *Retz.*, Observ. Bot. 4: 10 (1786).
Mascarenes, SE Asia, Melanesia, northern Australia; coastal sands.
MATERIALS: *Other materials/chemicals*: tubers used to scent coconut oil, Tonga (H73); rhizomes used for scenting coconut oil, beaten up and mixed with oil for perfume, Polynesia (H74).
ENVIRONMENTAL USES: *Erosion control*: sand binder (Burkill 1935).

Cyperus strigosus *L.*, Sp. Pl. 1: 47 (1753).
U.S.A. and Canada; margins of water bodies.

ANIMAL FOOD: *Unspecified part*: medium potential for grazing (USDA-NRCS 1999). WEEDS: *Unspecified weed*: (USDA-ARS 2000).

Cyperus submacropus *Kük.* in Engl., Pflanzenr. 4 (20), 101 Heft: 561 (1936). Tropical Africa; seasonally wet areas in grasslands.
ANIMAL FOOD: *Unspecified part*: Sudan (H236); grazed by all domestic livestock, Kenya (H272).

Cyperus subumbellatus *Kük.* in Engl., Pflanzenr. 4 (20), 101 Heft: 523 (1936). Tropical Africa, Mascarenes, W Indies; open forest, grasslands, seasonally wet areas.
FOOD: *Rhizomes*: edible (Heywood 1993); *Culms*: base edible after cooking (Burkill 1985; Peters 1992).
FOOD ADDITIVES: *Rhizomes*: rhizome aromatic and used as a food-flavouring (Burkill 1985).
ANIMAL FOOD: *Unspecified part*: culms eaten by domestic livestock (Burkill 1985).
MATERIALS: *Fibres*: thatching (Burkill 1985), Ghana (Abbiw 1990); brush making (Burkill 1985); *Other materials/chemicals*: rhizomes used as scrubbing brushes, Ghana (Abbiw 1990); rhizome scented (H210).
MEDICINES: *Infections/infestations*: swollen culm bases used to treat gonorrhoea, Nigeria (Burkill 1985); ground rhizomes used to treat ringworm, skin diseases and parasitic affections, Ghana (Abbiw 1990); *Injuries*: chewed culm is bandaged on a cut or wound for at least three days and replaced until healed, Nigeria (Burkill 1985).
ENVIRONMENTAL USES: *Ornamentals*: sold as 'Cyperus sumula' (D. A. Simpson pers. comm. 2000).
WEEDS: *Unspecified weed*: (Abbiw 1990); *Cultivation*: (Burkill 1985); *Waste places*: (Burkill 1985); *Gardens*: (Burkill 1985).

Cyperus tenellus *L. f.*, Suppl.: 103 (1782). S Africa, Australia and New Zealand; cultivated areas, grasslands, waste places, margins of water bodies.
WEEDS: *Cultivation*: (Healy & Edgar 1980); *Waste places*: (Healy & Edgar 1980); *Disturbed land*: (Healy & Edgar 1980); *Aquatic biotopes*: (Healy & Edgar 1980); *Irrigation ditches*: (Healy & Edgar 1980); *Gardens*: (Healy & Edgar 1980).

Cyperus tenuiculmis *Boeck.*, Linnaea 36: 286 (1870). Old World tropics (India, type), extending north to S Japan; open grasslands, grassy waysides, open forest.
MEDICINES: *Unspecified medicinal disorders*: unspecified use as a children's medicine, Tanzania (Burkill 1985).
WEEDS: *Unspecified weed*: Sierra Leone (H164); *Cultivation*: Nigeria (Burkill 1985); Gambia (H163); *Grassland*: Nigeria (Burkill 1985); *Gardens*: Nigeria (Burkill 1985).

Cyperus tenuis *Sw.*, Prodr.: 20 (1788). West tropical Africa, tropical America; open grasslands.

WEEDS: *Unspecified weed*: (Abbiw 1990); *Cultivation*: Cameroun (Burkill 1985); *Gardens*: Cameroun (H222).

Cyperus tenuispica *Steud.*, Syn. Pl. Glumac. 2: 11 (1855).
Old World tropics and subtropics; open, wet places.
     WEEDS: *Cultivation*: Togo (H203); *Rice fields*: (Kern 1974; Simpson & Koyama 1998; H201, H202).

Cyperus textilis *Thunb.*, Prodr. Pl. Cap. 1: 18 (1794).
S Africa; swamps and stream sides.
     MATERIALS: *Fibres*: basket-making sleeping mats, beehive houses, collecting baskets, winnowing baskets, grinding mats, rolled twine, cutting boards, S Africa (Burkill 1935; Kükenthal 1935 – 1936; Van Eyk & Gericke 2000); craftwork resource, has been planted in paddy fields, S Africa (Heinsohn 1990).

Cyperus tomaiophyllus *K. Schum.* in Engl., Pflanzenw. Ost-Afrikas C.: 122 (1895).
E tropical Africa; bogs and other wet habitats often in montane areas.
     ANIMAL FOOD: *Unspecified part*: grazed by all domestic livestock, Kenya (H268).
     MATERIALS: *Fibres*: used for thatching, Kenya (H268).

Cyperus usitatus *Burch.*, Trav. S. Africa 1: 417 (1822).
Southern Africa; seasonally wet habitats, grasslands.
     FOOD: *Tubers*: eaten raw (Peters 1992); roasted or boiled, Namibia, S Africa (Kükenthal 1935 – 1936; Van Eyk & Gericke 2000).
     ANIMAL FOOD: *Tubers*: (Fox & Norwood Young 1982); *Rhizomes*: (Kükenthal 1935 – 1936).
     MATERIALS: *Other materials/chemicals*: Oil from tuber used as a scented pomade, Botswana (H275).

Cyperus ustulatus *A. Rich.*, Essai Fl. N.Z.: 101, t. 17 (1832).
New Zealand; damp places, margins of water bodies, gullies and seepages.
     ENVIRONMENTAL USES: *Ornamentals*: (Grounds 1989).
     WEEDS: *Pasture*: (Healy & Edgar 1980).

Cyperus vaginatus *R. Br.*, Prodr.: 213 (1810).

subsp. **gymnocaulos** (*Steud.*) *Kük.* in Engl., Pflanzenr. 4 (20), 101 Heft: 190 (1936).
Cyperus gymnocaulos Steud., Syn. Pl. Glumac. 2: 12 (1855).
Australia; margins of rivers and creeks, on sand and sandy clay.
     ANIMAL FOOD: *Unspecified part*: fodder, Australia (H154).

Cyperus victoriensis *C. B. Clarke*, Kew Bull. Addit. Ser. 8: 12 (1908).
Australia; open grassy areas on clay.
     FOOD: *Rhizomes*: vegetable and famine food, Australia (Anon. 1994 – 2000).

Cyperus zollingeri *Steud.*, Syn. Pl. Glumac. 2: 17 (1855).

Tropical Africa to Australia; open, seasonally wet areas.

ANIMAL FOOD: *Unspecified part*: grazing for cattle, Senegal (Burkill 1985).

WEEDS: *Perennial crops*: sugar cane, Zimbabwe (H276).

DESMOSCHOENUS *Hook. f.*

Desmoschoenus spiralis (*A. Rich.*) *Hook. f.*, Fl. Nov.-Zel. 1: 272 (1853).
Isolepis spiralis A. Rich. in Dumont d' Urville, Voy. Astrolabe: 105, t. 19 (1832).
New Zealand; coastal, on sand dunes.

ANIMAL FOOD: *Unspecified part*: heavily grazed by rabbits and hares (Partridge 1992).

MATERIALS: *Fibres*: used by Maoris to bind and colour artefacts made from *Phormium tenax*, New Zealand (Grounds 1989; Moore & Edgar 1970).

ENVIRONMENTAL USES: *Erosion control*: sand binder (Grounds 1989; Moore & Edgar 1970; Partridge 1992); *Revegetators*: after mining (Partridge 1992); *Ornamentals*: (Grounds 1989).

DIPLACRUM *R. Br.*

Diplacrum caricinum R. Br., Prodr.: 241 (1810).
Scleria caricina (R. Br.) Benth., Fl. Austral. 7: 426 (1878).
India and Sri Lanka to N Australia also SE China and Japan; open wet grassland, savannahs, swamps and rice fields.

WEEDS: *Rice fields*: Thailand (Simpson & Koyama 1998).

ELEOCHARIS *R. Br.*

Eleocharis acicularis (*L.*) *Roem. & Schult.*, Syst. Veg. 2: 154 (1817).
Scirpus acicularis L., Sp. Pl. 1: 48 (1753).
Widespread in the northern hemisphere; open, damp places.

ENVIRONMENTAL USES: *Ornamentals*: (Huxley 1992).

WEEDS: *Grassland*: (Kühn 1982); *Aquatic biotopes*: (Kühn 1982); *Rice fields*: (Kühn 1982).

Eleocharis acutangula (*Roxb.*) *Schult.* in Roem. & Schult., Mant. 2: 91 (1824).
Scirpus acutangulus Roxb., Fl. Ind. 1: 216 (1820).
Pantropical; open, wet places.

MATERIALS: *Fibres*: matting, Borneo, Sumatra (Burkill 1935; H319); provides material for weaving, Brazil (Pio Corrêa 1926).

WEEDS: *Rotation crops*: (Kühn 1982); *Aquatic biotopes*: (Kühn 1982); *Rice fields*: (Kühn 1982; Simpson & Koyama 1998).

Eleocharis atropurpurea *C. Presl* in J. Presl & C. Presl, Reliq. Haenk. 1: 196 (1828).
Pantropical; open, wet places.

WEEDS: *Aquatic biotopes*: (Kühn 1982).

Eleocharis calva *Torr.*, Fl. New York 2: 346 (1843).
N America, Pacific Is; open wet places.
MATERIALS: *Fibres*: red basal sheath incorporated into mats, Niihau, U.S.A. (Hawaii) (Jobe 1991).

Eleocharis capillacea *Kunth*, Enum. Pl. 2: 139 (1837).
Venezuela, Brazil, Paraguay; river banks, swampy meadows.
ANIMAL FOOD: *Unspecified part*: forage, Brazil (Pio Corrêa 1926).

Eleocharis congesta *D. Don*, Prodr. Fl. Nepal: 41 (1825).
Eleocharis pellucida C. Presl in J. Presl & C. Presl, Reliq. Haenk. 1: 196 (1828).
India to China and Japan, W and C Malesia; shallow water, swampy places, ditches.
WEEDS: *Aquatic biotopes*: (Kühn 1982); *Rice fields*: (Kühn 1982; Simpson & Koyama 1998).

Eleocharis dulcis (*Burm. f.*) *Hensch.*, Vita Rumph.: 186 (1833).
Andropogon dulce Burm. f., Fl. Ind.: 219 (1768).
Pantropical; cultivated in Asia; open marshy places along coasts and inland.
FOOD: *Corms*: edible raw and cooked, palatable and nutritious, widely eaten, China, India, Japan, Philippines, U.S.A., Vietnam (Burkill 1985; Nguyen 1993; Huxley 1992); flour made from corms, N China, commercial starch can be prepared and fresh or canned corms exported (Burkill 1935; Burkill 1985; Kern 1974; Tim *et al.* 1983).
ANIMAL FOOD: *Leaves/culms/aerial parts*: promising plant for leaf protein concentrate extraction used as cattle fodder, India (Pandey & Srivastava 1991).
MATERIALS: *Fibres*: cultivated for matting, Sumatra, N Sulawesi (Burkill 1935; Burkill 1985); culms used for basket making (Heywood 1993); sometimes cultivated for making grass skirts, Papua New Guinea (Kern 1974; Leach & Osborne 1985).
MEDICINES: *Unspecified medicinal disorders*: tubers considered cooling and sometimes used in jaundice (Nguyen 1993), China (J. C. Shaw, C. S. Tong & L. Wong pers. comm. 1995); *Infections/infestations*: sap expressed from corms (puchiine) has antibiotic action (Burkill 1985; Nguyen 1993).
ENVIRONMENTAL USES: *Ornamentals*: (Huxley 1992).
WEEDS: *Aquatic biotopes*: (Kühn 1982; Pandey & Srivastava 1991); *Rice fields*: (Kühn 1982).

Cultivated material is usually referable to var. *tuberosa* (Roxb.) T. Koyama.

Eleocharis elegans (*Kunth*) *Roem. & Schult.*, Syst. Veg. 2: 150 (1817).
Scirpus elegans Kunth in Humb., Bonpl. & Kunth, Nov. Gen. Sp. 1: 226 (1816).
Costa Rica to Argentina. Caribbean Is.; wet or flooded areas.
MATERIALS: *Fibres*: leaves, matting (Anon. 1994 – 2000).
WEEDS: *Aquatic biotopes*: (Kühn 1982); *Rice fields*: (Kühn 1982).

Eleocharis filiculmis *Kunth*, Enum. 2: 144 (1837).
America: Mexico to Argentina; river banks, swamps, alluvial flats, savannahs, forests.
WEEDS: *Aquatic biotopes*: (Kühn 1982).

Eleocharis flavescens (*Poir.*) *Urb.*, Symb. Antill. 4: 116 (1903).
Scirpus flavescens Poir. in Lam., Encycl. 6: 756 (1805).
Southern U.S.A. to Brazil; moist places.
    ANIMAL FOOD: *Unspecified part*: forage, Brazil (Pio Corrêa 1926).

Eleocharis geniculata (*L.*) *Roem. & Schult.*, Syst. Veg. 2: 150 (1817).
Scirpus geniculatus L., Sp. Pl. 1: 48 (1753).
Tropical, subtropical and warm-temperate regions worldwide; wet, grassy places
including rice fields.
    ANIMAL FOOD: *Unspecified part*: forage, Brazil (Pio Corrêa 1926).
    MATERIALS: *Cane/reed*: culms useful for making packsaddles, Venezuela (Pittier
1971).
    MEDICINES: *Unspecified medical disorders*: infusion of rhizome has tonic properties,
Venezuela (Pittier 1971).
    WEEDS: *Grassland*: (Kühn 1982); *Aquatic biotopes*: (Kühn 1982); *Rice fields*: (Kühn
1982).

Eleocharis interstincta (*Vahl*) *Roem. & Schult.*, Syst. Veg. 2: 149 (1817).
Scirpus interstinctus Vahl, Enum. Pl. 2: 251 (1806).
Southern U.S.A. to Brazil and Bolivia, W Indies; swampy areas.
    MATERIALS: *Fibres*: has been used for chair seating, Dominican Republic (H320).

Eleocharis mutata (*L.*) *Roem. & Schult.*, Syst. Veg. 2: 155 (1817).
Scirpus mutatus L., Syst. Nat 10, 2: 867 (1759).
C America, S America to Paraguay, W Indies, W Africa; swampy areas, often in
brackish water.
    ANIMAL FOOD: *Unspecified part*: forage, Brazil (Pio Corrêa 1926).
    MATERIALS: *Cane/reed*: weaving, Brazil (Pio Corrêa 1926).
    WEEDS: *Rice fields*: marginal coastal rice-padis and cleared mangrove areas
(Burkill 1985).

Eleocharis ochrostachys *Steud.*, Syn. Pl. Glumac. 2: 80 (1855).
S and SE Asia, Pacific Is.; open wet places.
    MATERIALS: *Fibres*: sometimes used for making bags, Sumatra, Borneo (Kern 1974).

Eleocharis pallens *S. T. Blake*, Proc. Roy. Soc. Queensland 1937, 49: 154 (1938).
Australia; wet and flooded areas.
    ANIMAL FOOD: *Leaves/culms/aerial parts*: culms grazed (Anon. 1994 – 2000).

Eleocharis palustris (*L.*) *Roem. & Schult.*, Syst. Veg. 2: 151 (1817).
Scirpus palustris L., Sp. Pl. 1: 47 (1753).
Temperate to subtropical regions worldwide; margins of standing water, marshy
areas.
    ENVIRONMENTAL USES: *Ornamentals*: (Huxley 1992).
    WEEDS: *Aquatic biotopes*: (Kühn 1982).

Eleocharis parvula (*Roem. & Schult.*) *Link ex Bluff, Nees & Schauer*, Comp. Fl. German. 2. 1 (1): 93 (1836).
Scirpus parvulus Roem. & Schult., Syst. Veg. 2: 124 (1817).
Europe, N & S Africa, Japan, N America; wet, muddy places often in coastal areas or saline areas by inland lakes.
  ENVIRONMENTAL USES: *Ornamentals*: (Huxley 1992).

Eleocharis philippinensis *Svenson*, Rhodora 31: 155, f. 9 (1929).
Indo-China, Hainan, Malesia (Philippines, type), N Australia and New Caledonia; open, swampy places and rice fields at low altitudes.
  WEEDS: *Rice fields*: (Simpson & Koyama 1998).

Eleocharis pusilla *R. Br.*, Prodr.: 225 (1810).
Australia and New Zealand; damp sandy shores of lakes and streams, sand hollows among dunes.
  ENVIRONMENTAL USES: *Ornamentals*: (Grounds 1989).

Eleocharis retroflexa (*Poir.*) *Urb.*, Symb. Antill. 2: 165 (1900).
Scirpus retroflexus Poir. in Lam., Encycl. 6: 753 (1805).

subsp. **chaetaria** (*Roem. & Schult.*) *T. Koyama*, Bull. Natl. Sci. Mus. Tokyo 17: 68, f. 2 (1974).
Eleocharis chaetaria Roem. & Schult., Syst. Veg. 2: 154 (1817).
India (type), Sri Lanka, Indo-China, Malesia and Australia; wet grassy places, margins of ponds and ditches, rice fields.
  WEEDS: *Rice fields*: (Simpson & Koyama 1998).

Eleocharis sphacelata *R. Br.*, Prodr.: 224 (1810).
New Guinea, Australia and New Zealand; swamps, lake margins.
  FOOD: *Rhizomes*: eaten raw and cooked (Anon. 1994 – 2000).
  ANIMAL FOOD: *Leaves/culms/aerial parts*: culm grazed by cattle, sheep, goats (Anon. 1994 – 2000).
  MATERIALS: *Fibres*: used to make rush skirts (Kern 1974); sometimes cultivated for making grass skirts, Papua New Guinea (Leach & Osborne 1985).

Eleocharis spiralis (*Rottb.*) *Roem. & Schult.*, Syst. Veg. 2: 155 (1817).
Scirpus spiralis Rottb., Descr. Icon. Rar. Pl.: 45, t. 15, f. 1 (1773).
Tropical Africa, Mascarenes, India to S China, Malesia, Australia; open wet places, pools, swamps.
  MATERIALS: *Fibres*: matting, Java (Kern 1974).

Eleocharis vivipara *Link*, Hort. Berol. 1: 283 (1827).
SE U.S.A.; Ditches, muddy places.
  ENVIRONMENTAL USES: *Ornamentals*: (Huxley 1992).

ERIOPHORUM *L.*

Eriophorum angustifolium *Honck.*, Verz. Gew. Teutschl.: 153 (1782).
Europe except Meditteranean and SE, N America, Arctic; wet peaty areas or shallow water.
    MATERIALS: *Fibres*: silky, elongated perianth-segments used to make paper and candle wicks and stuff pillows, U.K. (Pratt 1900); *Other materials/chemicals*: (Pratt 1900).
    ENVIRONMENTAL USES: *Ornamentals*: (Grounds 1989; Huxley 1992).

Eriophorum chamissonis *C. A. Mey.*, Mém. Acad. Imp. Sci. St.-Pétersbourg Divers Savans 1: 204, t. 3 (1831).

var. **albidum** (*Nyl.*) *Fernald*, Rhodora 7 (77): 84 (1905).
Eriphorum russeolum Fries var. *albidum* Nyl., Bot. Not. 1857 (4): 58 (1857).
Western N America; bogs, pool margins, tundra.
    ENVIRONMENTAL USES: *Ornamentals*: (Huxley 1992).

Eriophorum comosum *Nees* in Wight, Contr. Bot. India: 110 (1834).
India, Himalayan region; rocky crevices.
    MATERIALS: *Fibres*: much used for rope-bridges, India (Hooker 1857).

Eriophorum latifolium *Hoppe*, Bot. Taschenb.: 108 (1800).
N America, most of Europe, W Asia, Siberia; bogs and marshes.
    ENVIRONMENTAL USES: *Ornamentals*: (Grounds 1989; Huxley 1992).

Eriophorum scheuchzeri *Hoppe*, Taschenb.: 104 (1800).
N America, N Europe, mountains of C and S Europe, Siberia; bogs, wet tundra, muddy or peaty lake margins.
    ENVIRONMENTAL USES: *Ornamentals*: (Huxley 1992).

Eriophorum vaginatum *L.*, Sp. Pl. 1: 52 (1753).
N temperate regions; bogs and moorland.
    MATERIALS: *Fibres*: elongated and cotton-like perianth-segments used to make yarn and fabric, U.K. (H302 — Fig. 3F, H303 — Fig. 3G).
    FUELS: *Tinder*: perianth-segments have been used as tinder, Canada (H309).
    ENVIRONMENTAL USES: *Ornamentals*: (Grounds 1989; Huxley 1992).

Eriophorum viridi-carinatum (*Engelm.*) *Fernald* in Britton & Brown, Ill. Fl. N. U.S. 2, 1: 325 (1913).
Eriophorum latifolium Hoppe var. *viridi-carinatum* Engelm., Amer. J. Sci. 46: 103 (1844).
N America; Peaty swamps, ditches, marshes.
    ENVIRONMENTAL USES: *Ornamentals*: (Huxley 1992).

FICINIA *Schrad.*

Ficinia nodosa (*Rottb.*) *Goetgh., Muasya & D. A. Simpson*, Novon 10 (2): 132 – 133 (2000).

Scirpus nodosus Rottb., Descr. Icon. Rar. Pl.: 52. t. 8, f. 3 (1773).
Isolepis nodosa (Rottb.) R. Br., Prodr. 221 (1810).
Widespread in the southern hemisphere; coastal sand dunes and streamsides.
   MATERIALS: *Cane/reed*: culms used to make packsaddles, Venezuela (Pittier 1971).

FIMBRISTYLIS *Vahl*

Fimbristylis acuminata *Vahl*, Enum. Pl. 2: 285 (1806).
India and Sri Lanka, through Malesia to N Australia and north-eastwards to S China and the Ryukyu Is.; open, wet or muddy places.
   WEEDS: *Aquatic biotopes*: (Kühn 1982); *Rice fields*: (Kühn 1982).

Fimbristylis aestivalis (*Retz.*) *Vahl*, Enum. Pl. 2: 288 (1806).
Scirpus aestivalis Retz., Observ. Bot. 4: 12 (1786).
India to northern Australia; open damp places, swamps, rice fields.
   MEDICINES: *Inflammation*: may be used in poulticing, along with *Cassia alata* L. (*Leguminosae*), Malaysia (Burkill 1935).
   WEEDS: *Rotation crops*: (Kühn 1982); soya bean fields, Thailand (Radanachaless & Maxwell 1994); *Perennial crops*: (Kühn 1982); *Aquatic biotopes*: (Kühn 1982); *Rice fields*: (Kühn 1982; Simpson & Koyama 1998).

Fimbristylis autumnalis (*L.*) *Roem. & Schult.*, Syst. Veg. 2: 97 (1817).
Scirpus autumnalis L., Mant. Pl. 2: 180 (1771).
N and S America, western, central and southern Africa, India, Japan; wet places, wet cultivated areas.
   WEEDS: *Unspecified weed*: potential seed contaminant (USDA-ARS 2000); *Rotation crops*: (Kühn 1982); *Aquatic biotopes*: (Kühn 1982); *Rice fields*: (Kühn 1982).

Fimbristylis bisumbellata (*Forssk.*) *Bubani*, Dodecanthea 30 (1850).
Scirpus bisumbellatus Forssk., Fl. Aegypt.-Arab. 1: 15 (1775).
Warm temperate to tropical regions of the Old World; wet places.
   WEEDS: *Cultivation*: (Täckholm & Drar 1950); *Rice fields*: (Simpson & Koyama 1998).

Fimbristylis complanata (*Retz.*) *Link*, Hort. Berol. 1: 292 (1833).
Scirpus complanatus Retz., Observ. Bot. 5: 14 (1789).
Pantropical; wet grassland, rice fields, seashores and cultivated ground.
   MATERIALS: *Fibres*: used in basketry, Sri Lanka (H321).
   WEEDS: *Rotation crops*: (Kühn 1982); *Aquatic biotopes*: (Kühn 1982); *Rice fields*: (Simpson & Koyama 1998).

Fimbristylis cymosa *R. Br.*, Prodr.: 228 (1810).
Pantropical; open sandy or muddy areas.
   ENVIRONMENTAL USES: *Revegetators*: pioneer, covering and stabilising raw surfaces (Fosberg 1988).

Fimbristylis dichotoma (*L.*) *Vahl*, Enum. Pl. 2: 287 (1806).
Scirpus dichotomus L., Sp. Pl. 1: 50 (1753).
Pantropical; grassy waysides, cultivated ground, margins of rice fields, also in plantations, swamps and savannahs.

ANIMAL FOOD: *Unspecified part*: grazed by cattle especially the young plant (Burkill 1935; Burkill 1985); *Leaves/culms/aerial parts*: leaves, forage with sufficient food-value (Kern 1974).

MATERIALS: *Unspecified material type*: rhizomes aromatic, collected for this, India (Burkill 1985); *Fibres*: culms used for inferior matting, India, Philippines, Singapore (Burkill 1935; Burkill 1985; Kern 1974).

MEDICINES: *Skin/subcutaneous cellular tissue disorders:* used by the Shipibo-Conibo to maintain and encourage hair growth, mixed with *Genipa americana* L. (*Rubiaceae*) and rubbed on to the hair, Peru (Tournon *et al.* 1986).

ENVIRONMENTAL USES: *Soil improvers*: ploughed in as green manure (Burkill 1935; Burkill 1985).

WEEDS: *Unspecified weed*: Ghana (Abbiw 1990); *Rotation crops*: (Kühn 1982); soya bean fields, Thailand (Radanachaless & Maxwell 1994); *Perennial crops*: (Kühn 1982); *Grassland*: (Kühn 1982); *Waste places*: (Kühn 1982); *Aquatic biotopes*: (Kühn 1982); *Rice fields*: (Simpson & Koyama 1998).

Fimbristylis dura (*Zoll. & Moritzi*) *Merr.*, Philipp. J. Sci. 11: 53 (1916).
Isolepis dura Zoll. & Moritzi in Moritzi, Syst. Verz.: 97 (1846).
India and Sri Lanka, through Indo-China to Java and Borneo; open forest, more rarely in rice fields and along river banks.

MEDICINES: *Pregnancy/birth/puerperium disorders*: (Kern 1974); used after childbirth, Malaysia (Burkill 1935).

Fimbristylis falcata (*Vahl*) *Kunth*, Enum. Pl. 2: 239 (1837).
Scirpus falcatus Vahl, Enum. Pl. 2: 275 (1806).
India and Sri Lanka to Thailand, Indo-China, Philippines, New Guinea; grasslands.

MEDICINES: *Infections/infestations*: Santals use rhizome to relieve dysentery, India (Cauis & Banby 1935).

Fimbristylis ferruginea (*L.*) *Vahl*, Enum. Pl. 2: 291 (1806).
Scirpus ferrugineus L., Sp. Pl. 1: 50 (1753).
Pantropical; margins of brackish swamps and open, wet areas of clay or sandy soil by coast.

MATERIALS: *Fibres*: culms beaten to soften fibres and plaited into screens for huts, Somalia (Burkill 1985).

ENVIRONMENTAL USES: *Erosion control*: contributes to stabilisation of sandy areas (Burkill 1985).

WEEDS: *Waste places*: (Kühn 1982); *Aquatic biotopes*: (Kühn 1982).

Fimbristylis miliacea (*L.*) *Vahl*, Enum. Pl. 2: 287 (1806).
Scirpus miliaceus L., Syst. Nat. 10: 868 (1759).
Fimbristylis littoralis Gaudich. in Freyc., Voy. Uranie: 413 (1829).

Tropical and subtropical regions worldwide; wet places.

ANIMAL FOOD: *Unspecified part*: cattle (Burkill 1935; Burkill 1985; Kern 1974); grazed by all livestock, Sudan (Sudan Ministry 1980).

SOCIAL USES: *Miscellaneous social uses*: crushed rhizome used as an aphrodisiac by the Shipibo-Conibo, Peru (Tournon *et al.* 1986).

MEDICINES: *Unspecified medicinal disorders*: China (Burkill 1935); *Infections/infestations*: leaves used for poulticing in fever, Malaysia (Burkill 1935).

ENVIRONMENTAL USES: *Soil improvers*: ploughed in as green manure (Burkill 1935).

WEEDS: *Cultivation*: maize fields, India (Dangol 1992); *Rotation crops*: (Kühn 1982); *Perennial crops* (Kühn 1982); soya bean fields, Thailand (Radanachaless & Maxwell 1994); *Aquatic biotopes*: (Kühn 1982); *Rice fields*: (Burkill 1985; Kühn 1982; Simpson & Koyama 1998).

Fimbristylis nutans (Retz.) Vahl, Enum. Pl. 2: 285 (1806).
Scirpus nutans Retz., Observ. Bot. 4: 12 (1786).
India and Sri Lanka, Indo-China, Malesia (Malay Peninsula, type) to N Australia, S China and the Ryukyu Is.; wet grasslands, forest floors, rice fields and occasionally brackish marshes.

WEEDS: *Rice fields*: (Simpson & Koyama 1998).

Fimbristylis ovata (*Burm. f.*) *J. Kern*, Blumea 15: 126 (1967).
Carex ovata Burm. f., Fl. Ind.: 194 (1768).
Pantropical; dry or wet open grasslands, open forests, coastal rock outcrops.

ANIMAL FOOD: *Unspecified part*: grazing, all livestock, favoured by horses, Kenya (Burkill 1985).

MEDICINES: *Muscular-skeletal system disorders*: flowering culms plaited for bangles on wrists to relieve rheumatism, Africa (Burkill 1985).

Fimbristylis pauciflora R. Br., Prodr.: 225 (1810).
Myanmar through Indo-China to the Ryukyu Is., Malesia to the Caroline Is. and Australia; wet grasslands, savannahs, forest floors and swamp margins.

MEDICINES: *Pregnancy/birth/puerperium disorders*: rubbed on stomach to induce labour, Brunei (Simpson 1996).

ENVIRONMENTAL USES: *Soil improvers*: ploughed in as green manure, Malay Peninsula (Burkill 1935).

WEEDS: *Rice fields*: (Burkill 1935; Kern 1974).

Fimbristylis schoenoides (*Retz.*) *Vahl*, Enum. Pl. 2: 286 (1806).
Scirpus schoenoides Retz., Observ. Bot. 5: 14 (1789).
India, through Malesia to N Australia, also SE China and Taiwan; wet open grasslands, rice fields and on wet forest floors.

ENVIRONMENTAL USES: *Soil improvers*: ploughed in as green manure (Burkill 1935; Burkill 1985; Kern 1974).

WEEDS: *Rice fields*: (Burkill 1935; Burkill 1985; Kern 1974, Simpson & Koyama 1998).

Fimbristylis spadicea *Vahl*, Enum. Pl. 2: 294 (1806).
Tropical America; saline flatlands.
MATERIALS: *Fibres*: used to make paper, Mexico (H314).

Fimbristylis squarrosa *Vahl*, Enum. Pl. 2: 289 (1806).
Pantropical but excluding Malesia; open wet places.
MEDICINES: *Respiratory system disorders*: decoction of the plant is applied to obtain relief from sore throat, Nepal (Manandhar 1989).

Fimbristylis umbellaris (*Lam.*) *Vahl*, Enum. Pl. 2: 291 (1806).
Scirpus umbellaris Lam., Tabl. Encycl. 1: 141 (1791).
Fimbristylis globulosa (Retz.) Kunth, Enum. Pl. 2: 231 (1837).
India, Indo-China, S China, Ryukyu Is., Malesia, Micronesia and Polynesia; open swamps, wet grassy places, rice fields.
MATERIALS: *Fibres*: important for undyed, bleached and dyed hats from wild sources, Philippines (Burkill 1935); excellent raw material for floor mats, handbags, wall decorations slippers and cushions, good dye-absorbing characteristics, semi-flat culm for easy workability and strength for quality and durability, Philippines (Calanog & Reyes 1989); cultivated and used for making hats, baskets, bags, mats and string, Peninsular Malaysia, Sumatra, Java, Sulawesi (Burkill 1935; Kern 1974); plants dried and used as weaving material for hard-wearing mats (H280).
MEDICINES: *Blood system disorders*: medicine for splenomegaly, Philippines (Altschul 1973).
ENVIRONMENTAL USES: *Soil improvers*: ploughed in as green manure (Burkill 1935).
WEEDS: *Aquatic biotopes*: (Kühn 1982); *Rice fields*: (Simpson & Koyama 1998).

FUIRENA *Rottb.*

Fuirena ciliaris (*L.*) *Roxb.*, Hort. Bengal.: 81 (1814); Fl. Ind. 1: 184 (1820).
Scirpus ciliaris L., Mant. Pl. 2: 182 (1771).
Widely distributed in the Old World tropics and subtropics; open, wet or swampy places.
ANIMAL FOOD: *Unspecified part*: readily eaten by elephants, Kenya (Burkill 1985).
WEEDS: *Rotation crops*: soya bean fields, Thailand (Radanachaless & Maxwell 1994); *Grassland*: (Kühn 1982); *Aquatic biotopes*: (Kühn 1982); *Rice fields*: (Simpson & Koyama 1998).

Fuirena stricta *Steud.*, Syn. Pl. Glumac. 2: 128 (1855).

var. **chlorocarpa** (*Ridl.*) *Kük.* in Fries & Fries, Notizbl. Bot. Gart. Berlin-Dahlem 9: 310 (1925).
Fuirena chlorocarpa Ridl., Trans. Linn. Soc. London, Bot. 2: 159 (1884).
Sub-Saharan Africa; swampy ground, bogs.
WEEDS: *Rice fields*: (Burkill 1985).

Fuirena umbellata *Rottb.*, Descr. Icon. Rar. Pl.: 70, t. 19, f. 3 (1773).

Pantropical; areas of permanent shallow water, swamps, pools, ditches, marshy lake shores.

FOOD: *Tubers*: edible, Papua New Guinea (Altschul 1973).

FOOD ADDITIVES: *Unspecified part*: burnt to obtain salt from the ashes, Liberia (Burkill 1985); ashes used as a salt substitute, Ghana (Abbiw 1990).

ANIMAL FOOD: *Unspecified part*: livestock grazing, Ghana (Abbiw 1990; Burkill 1985).

MEDICINES: *Unspecified medicinal disorders*: new-born babies washed in an infusion of the leaves, Gambia (Burkill 1985).

ENVIRONMENTAL USES: *Erosion control*: may serve as a mud-binder to resist tidal scouring (Burkill 1985); *Soil improvers*: ploughed in as green manure (Burkill 1935).

WEEDS: *Grassland*: (Kühn 1982); *Aquatic biotopes*: (Kühn 1982); *Rice fields*: (Burkill 1985).

GAHNIA *J. R. Forst. & G. Forst.*

Gahnia aspera (*R. Br.*) *Spreng.*, Syst. Veg. 2: 114 (1825).
Lampocarya aspera R. Br., Prodr.: 238 (1810).
Maluku, New Guinea, N and E Australia, New Caledonia, Pacific Is.; open forests, coastal rocky areas.

ANIMAL FOOD: *Leaves/culms/aerial parts*: culms grazed, Australia (Anon. 1994 – 2000).

Gahnia grandis (*Labill.*) *S. T. Blake*, Contr. Queensland Herb. 8: 33 (1969).
Scleria grandis Labill., Voy. Rech. Pérouse: 146 (1800).
Australia; grassy heaths, wet sclerophyll forest.

FOOD: *Aerial parts*: leaf buds eaten, Australia (Irvine 1957).

Gahnia procera *J. R. Forst. & G. Forst.*, Char. Gen. Pl.: 52, t. 26 (1776).
New Zealand; mountain forest, bog or scrub.

ENVIRONMENTAL USES: *Ornamentals*: (Grounds 1989).

Gahnia trifida *Labill.*, Nov. Holl. Pl. 1: 89, t. 116 (1805).
Australia; open areas.

MATERIALS: *Fibres*: used to make brushes, Australia (H315).

Gahnia tristis *Nees* in Hook. & Arn., Bot. Beechey Voy. 228 (1837).
Macao (type), S China, Ryukyu Is., Indo-China, Malesia; dry places near coast, savannah grasslands, forest margins.

ANIMAL FOOD: *Inflorescences/infructescence/nutlets*: bird seed, China (J. C. Shaw, C. S. Tong & L. Wong pers. comm. 1995).

HYPOLYTRUM *Rich. ex Pers.*

Hypolytrum compactum *Nees & Meyen ex Kunth*, Enum. Pl. 2: 271 (1837).
Andaman Is., Indo-China, Malesia, N Australia; rainforests.

MEDICINES: *Unspecified medicinal disorders*: young culm chewed with betel and applied to ulcers, Philippines (Altschul 1973).

Hypolytrum heteromorphum *Nelmes*, Kew Bull. 10: 522 (1955).
W and C Africa, Uganda, Tanzania; swampy areas in evergreen forests.
    SOCIAL USES: '*Religious*' *uses*: decoction with plants of close affinity used as a face wash in cases of insanity, Congo (Burkill 1985).

Hypolytrum laxum Kunth, Enum. Pl. 2: 270 (1837).
Tropical S America; forest.
    MEDICINES: *Unspecified medicinal disorders*: considered emollient, stomachic and diuretic, Brazil (Pio Corrêa 1926).

Hypolytrum poecilolepis *Nelmes*, Kew Bull. 10: 77 (1955).
W Africa; wet evergreen forest.
    SOCIAL USES: '*Religious*' *uses*: twins of Mende women placed on a mat specially made of the leaves before they are washed, to bring good luck, Sierra Leone. (Burkill 1985).

Hypolytrum purpurascens *Cherm.*, Bull. Soc. Bot. France 80: 508 (1933).
W Africa south to Angola; forest, swampy places, often on margins of water.
    MATERIALS: *Fibres*: matting, Ghana (Abbiw 1990).
    MEDICINES: *Respiratory system disorders*: leaf decoction used as a cough medicine, Sierra Leone (Burkill 1985).

ISOLEPIS *R. Br.*

Isolepis cernua (*Vahl*) *Roem. & Schult.*, Syst. Veg. 2: 106 (1817).
Scirpus cernuus Vahl, Enum. Pl. 2: 245 (1806).
Widepread, but excluding tropical Africa and SE Asia; damp or wet places.
    MEDICINES: *Unspecified medicinal disorders*: S Africa (Cauis & Banby 1935).
    ENVIRONMENTAL USES: *Ornamentals*: (Grounds 1989; Huxley 1992).

Isolepis prolifera (*Rottb.*) *R. Br.*, Prodr.: 223 (1810).
Scirpus prolifer Rottb., Descr. Icon. Rar. Pl.: 55, t. 17, f. 2 (1773).
S Africa, Australia; damp, sandy areas.
    ENVIRONMENTAL USES: *Ornamentals*: (Huxley 1992).

Isolepis setacea (*L.*) *R. Br.*, Prodr.: 221 (1810).
Scirpus setaceus L., Sp. Pl. 1: 49 (1753).
Most of Europe, Africa, N and W Asia, Australia; damp places.
    ENVIRONMENTAL USES: *Ornamentals*: (Huxley 1992).

KOBRESIA *Willd.*

NOTE: Noltie (1993) states that in Bhutan, N India and Nepal 'although Kobresias are relatively humble in appearance, they are of considerable economic significance forming a major part of high pastures grazed in summer by yak and sheep'.

Kobresia capillifolia *C. B. Clarke*, J. Linn. Soc., Bot. 20: 378 (1883).
Himalayas; scrub, meadows and alpine turf.
    ANIMAL FOOD: *Unspecified part*: eaten by livestock, India (R. C. Srivastava pers. comm. 1994).

Kobresia pygmaea *C. B. Clarke* in Hook. f., Fl. Brit. India 6: 696 (1894).
Himalayas; Tibetan plateau; open grassy rocky and rocky areas at high altitude.
    ANIMAL FOOD: *Unspecified part*: a main fodder-plant, distributed in areas of Yak-pasturing, Tibet (Dickoré 1994); probably the most important plant of Tibet, highly adapted to grazing (Dickoré 1994).
    MATERIALS: *Unspecified material type*: sods used for building walls and rooves, Tibet (Dickoré 1994).
    ENVIRONMENTAL USES: *Erosion control*: Tibet (Dickoré 1994).

Kobresia schoenoides *Boeck.*, Linnaea 39: 7 (1875).
W Asia through Himalayas to W China; meadows, moors, flushes, fens at high altitude.
    ANIMAL FOOD: *Unspecified part*: eaten by livestock, India (R. C. Srivastava pers. comm. 1994).

KYLLINGA *Rottb.*

Kyllinga brevifolia *Rottb.*, Descr. Icon. Rar. Pl.: 13, t. 4, f. 3 (1773).
Cyperus brevifolius (Rottb.) Hassk., Cat. Hort. Bot. Bogor.: 24 (1844).
Pantropical and warm temperate regions; open damp, grassy areas, roadsides, waste places.
    ANIMAL FOOD: *Unspecified part*: eaten by sheep, cattle and horses (Burkill 1935; Healy & Edgar 1980); food value satisfactory, but yield scant (Kern 1974).
    MEDICINES: *Blood system disorders*: with leaves of Denyung (Cane grass), macerated in cold water and drunk for spleen trouble, New Hebrides (Altschul 1973); *Digestive system disorders*: leaves taken internally for diarrhoea (Burkill 1935); used for fistula, pustules, tumours and stomach and intestinal problems, India (Mukhopadhyay & Ghosh 1992); leaves crushed and juice taken for stomach problems, India (Kapur *et al.* 1992b); *Infections/infestations*: rhizome used for poulticing sores on legs, Malaysia (Cauis & Banby 1935, Burkill 1935); decoction of whole plant febrifuge, Philippines (Altschul 1973); used against malaria, China (J. C. Shaw, C. S. Tong & L. Wong pers. comm. 1995); *Inflammation*: whole plant used to relieve swelling, China (J. C. Shaw, C. S. Tong & L. Wong pers. comm. 1995); *Pain*: whole plant used to relieve pain, China (J. C. Shaw, C. S. Tong & L. Wong pers. comm. 1995); *Poisonings*: used to relieve snake-bite, China (J. C. Shaw, C. S. Tong & L. Wong pers. comm. 1995); *Pregnancy/birth/puerperium disorders*: with leaf of 'Naiwas tree' and 'Tomi-rirri' and 'Nesiv-nesip' (shrub) macerated in cold water for good health during pregnancy, New Hebrides (Altschul 1973); *Skin/subcutaneous*

cellular tissue disorders: (Burkill 1935; Cauis & Banby 1935); leaves crushed and juice is taken for skin problems, India (Kapur *et al.* 1992b).

WEEDS: *Cultivation*: palm plantations, Cameroun (Burkill 1985); *Rotation crops*: soya bean fields, Thailand (Radanachaless & Maxwell 1994); *Perennial crops*: (Kühn 1982); *Grassland*: (Kühn 1982); *Waste places*: (Kühn 1982); *Gardens*: (Healy & Edgar 1980).

var. **gracillima** *(Miq.) Kük.*, Acta Horti Gothob. 5: 107 (1929).
Kyllinga gracillima Miq., Ann. Mus. Bot. Lugduno-Batavum 2: 142 (1866).
Cyperus brevifolius (Rottb.) Hassk. var. *gracillimus* (Miq.) Kük. in Engl., Pflanzenr. 4 (20), 101 Heft: 603 (1936).
E Asia (China, Japan, Russia), Nepal; habitat not recorded.
WEEDS: *Unspecified weed*: (USDA-ARS 2000).

Kyllinga erecta *Schumach.* in Schumach. & Thonn., Beskr. Guin. Pl.: 42 (1827).
Cyperus erectus (Schumach.) Mattf. & Kük. in Engl., Pflanzenr. 4 (20), 101 Heft: 588 (1936).
Tropical Africa, Mascarenes; wet grasslands, margins of water bodies.
FOOD: *Culms*: base of culm, Africa (Peters 1992).
FOOD ADDITIVES: *Rhizomes*: aromatic rhizomes put in food and medicine as a flavouring, N Nigeria (Burkill 1985); *Exudates*: juice used as flavouring, Ghana (Abbiw 1990).
ANIMAL FOOD: *Unspecified part*: fodder, Ghana (Abbiw 1990).
MATERIALS: *Fibres*: plant bundles placed beneath animal skins being worked, making them supple, Lesotho (Burkill 1985); *Other materials/chemicals*: aromatic rhizomes sold in markets for use as fumigants, N Nigeria (Burkill 1985).
MEDICINES: *Infections/infestations*: plant boiled in palm-wine, infusion taken for use against prurient skin afflictions, Congo (Burkill 1985).
WEEDS: *Cultivation* (Burkill 1985); *Rotation crops*: (Kühn 1982); *Grassland*: (Kühn 1982); *Waste places* (Kühn 1982).

Kyllinga nemoralis *(J. R. & G. Forst.) Dandy ex Hutch. & Dalziel*, Fl. W. Trop. Afr. 2: 486, in key, & 487 (1936).
Kyllinga monocephala Rottb., Descr. Icon. Rar. Pl.: 13, t. 4, f. 4 (1773), nom. illegit.
Thryocephalon nemoralis J. R. Forst. & G. Forst., Char. Gen. Pl. 130 (1776).
Cyperus kyllingia Endl., Cat. Horti Vindob. 1: 94 (1842).
Pantropical but rare in S America and most frequent in tropical Asia; open to partly shaded waste places, grasslands, secondary forest.
ANIMAL FOOD: *Unspecified part*: eaten by cattle unless it is very old (Burkill 1935); eaten by cattle and horses, when grass is scarce, satisfactory food value, SE Asia (Kern 1974).
MEDICINES: *Unspecified medicinal disorders*: leaves used medicinally, China, India (Burkill 1935, H326); *Digestive system disorders*: rhizome stops diarrhoea, Amboinese (Burkill 1935); *Infections/infestations*: rhizome is a good refrigerant, much used in fevers, India (Cauis & Banby 1935; Mukhopadhyay & Ghosh 1992); rhizome used for treating measles, Sulawesi (Burkill 1935); *Inflammation*: inflorescence used for

poulticing gathered nails, China (Burkill 1935); *Poisonings*: rhizome used as an antidote, India, Indonesia (Burkill 1935; Cauis & Banby 1935; Mukhopadhyay & Ghosh 1992); *Skin/subcutaneous cellular tissue disorders*: crushed plant is used in folk medicines as a poultice for ulcers, Vietnam (Nguyen 1993).

WEEDS: *Unspecified weed*: common weed, Vietnam, Cambodia and Laos (Nguyen 1993); *Cultivation* (Burkill 1985); *Rotation crops*: (Kühn 1982); *Rotation crops*: soya bean fields, Thailand (Radanachaless & Maxwell 1994); *Perennial crops*: (Kühn 1982); *Grassland*: (Kühn 1982); *Waste places* (Kühn 1982).

Kyllinga odorata *Vahl*, Enum. Pl. 2: 382 (1806).
Cyperus sesquiflorus (Torr.) Mattf. & Kük. in Engl., Pflanzenr. 4(20), 101 Heft: 591, f. 6 E – J (1936).
Pantropical; open grasslands, disturbed ground.

ANIMAL FOOD: *Unspecified part*: horses and mules graze it without apparent ill-effect, Kenya (Burkill 1985).

MEDICINES: *Unspecified medicinal disorders*: Brazil (Cauis & Banby 1935); infusion of rhizome astringent and considered to be stomachic, Venezuela (Pittier 1971); decoction of whole plant used as an external bath for fevers by Wayãpi, French Guiana (Milliken 1997).

WEEDS: *Cultivation*: Ethiopia and E Africa (Burkill 1985); *Gardens*: lawns, Ethiopia and E Africa (Burkill 1985).

Kyllinga peruviana *Lam.*, Encycl. 3: 366 (1792).
Cyperus peruvianus (Lam.) F. N. Williams, Bull. Herb. Boiss., sér. 2, 7: 90 (1907).
W Africa, W Indies, central and tropical S America; coastal areas including dunes and foreshores.

MATERIALS: *Fibres*: stuffing for mattresses, Ghana (Burkill 1985).

Kyllinga polyphylla *Willd. ex Kunth*, Enum. Pl. 2: 134 (1837).
Cyperus aromaticus (Ridl.) Mattf. & Kük. in Engl., Pflanzenr. 4 (20), 101 Heft: 581 (1936).
Native to E Africa, introduced into SE Asia and Pacific Is.; open, often seasonally wet grassy places.

WEEDS: *Unspecified weed*: ditches and roadsides, Australia (Parsons & Cuthbertson 1992); *Cultivation*: root crops, coconut and sugar plantations, Australia (Parsons & Cuthbertson 1992); *Rice fields*: Australia (Parsons & Cuthbertson 1992); *Gardens*: lawns, Australia (Parsons & Cuthbertson 1992); *Pasture*: Australia (Parsons & Cuthbertson 1992).

var. **elatior** *Kük.*, Notizbl. Bot. Gart. Berlin-Dahlem 9: 300 (1925).
NE Africa; grasslands, particularly on margins of swamps and water bodies.
WEEDS: *Cultivation*: Ethiopia (Burkill 1985).

Kyllinga pumila *Michx.*, Fl. Bor.-Amer. 1: 28 (1803).
Cyperus densicaespitosus (Ridl.) Mattf. & Kük. in Engl., Pflanzenr. 4 (20), 101 Heft: 597 (1936).

Southern N America, S America, tropical Africa; margins of water bodies.

FOOD ADDITIVES: *Rhizomes*: rhizomes chewed and added as a flavouring to food or medicine (Burkill 1985); *Exudates*: juice used a flavouring, Ghana (Abbiw 1990).

ANIMAL FOOD: *Unspecified part*: fodder, Ghana (Abbiw 1990).

MATERIALS: *Other materials/chemicals*: rhizomes fragrant, used as a fumigant, Africa (Burkill 1985).

MEDICINES: *Unspecified medicinal disorders*: decoction of whole plant used as an external bath for fevers by Wayãpi, French Guiana (Milliken 1997).

WEEDS: *Cultivation*: (Svenson 1943).

Kyllinga squamulata *Thonn. ex Vahl*, Enum. Pl. 2: 381 (1806).
Cyperus metzii (Steud.) Mattf. & Kük. in Engl., Pflanzenr. 4 (20), 101 Heft: 612 (1936).
W Indies, tropical Africa to India and Indo-China; disturbed areas, waste places, often sandy.

FOOD ADDITIVES: *Culms*: swollen culm base chewed or added to food and medicine as flavour (Burkill 1985); *Exudates*: juice used as flavouring, Ghana (Abbiw 1990).

ANIMAL FOOD: *Unspecified part*: fodder, Ghana (Abbiw 1990).

MATERIALS: *Other materials/chemicals*: swollen culm base, sold as 'turare', used as a fumigant (Burkill 1985).

WEEDS: *Unspecified weed*: (Abbiw 1990); *Cultivation*: annual crops, Uganda, Tanzania (Burkill 1985).

Kyllinga triceps *Rottb.*, Descr. Icon. Rar. Pl.: 14, t. 4, f. 6 (1773).
Cyperus triceps (Rottb.) Endl., Cat. Horti Vindob. 1: 84 (1842).
Tropical Africa, Asia (rare in SE Asia), S China, Australia; damp grasslands, sides of water bodies.

MEDICINES: *Endocrine system disorders*: essential oil in rhizomes, used to promote the action of the liver, India (Cauis & Banby 1935); *Ill-defined symptoms*: useful in regulating the heat of the body, decoction of the rhizome relieves thirst in fever, India (Cauis & Banby 1935); *Infections/infestations*: bitter and cooling, good against infection, valuable in the treatment of blood diseases, decoction of the rhizomes relieves thirst in diabetes, India (Cauis & Banby 1935); *Injuries*: useful in healing wounds, India (Cauis & Banby 1935); *Nervous system disorders*: valuable in the treatment of nervous problems, India (Cauis & Banby 1935); *Poisonings*: bitter and cooling, good against poison, India (Cauis & Banby 1935); *Skin/subcutaneous cellular tissue disorders*: oil boiled with the rhizomes, used to relieve itching of the skin, India (Cauis & Banby 1935).

WEEDS: *Perennial crops*: young tea, Kenya (Burkill 1985); *Grassland*: E Africa (Burkill 1985); *Gardens*: lawns, E Africa (Burkill 1985).

Kyllinga welwitschii *Ridl.*, Trans. Linn. Soc. London, Bot. 2: 147 (1884).
Cyperus welwitschii (Ridl.) Lye in R. W. Haines & Lye, Sedges & Rushes E Afr. App. 3: 2 (1983).

Tropical Africa; damp soil near rivers and streams.

ANIMAL FOOD: *Unspecified part*: grazed by domestic livestock, Senegal (Burkill 1985).

LAGENOCARPUS *Nees*

Lagenocarpus adamantinus *Nees* in Mart., Fl. Bras. 2, 1: 165 (1842).
Tropical America; upland savannahs.

MATERIALS: *Other materials/chemicals*: source of industrial cellulose, Brazil ((Pio Corrêa 1926).

LEPIDOSPERMA *Labill.*

Lepidosperma gladiatum *Labill.*, Nov. Holl. Pl. 1: 15, t. 12 (1805).
Australia; coastal sandy areas and sand dunes.

MATERIALS: *Fibres*: paper, Australia (H300 — Fig. 3D).

FOOD: *Leaves*: blanched portion at base of inner leaves edible, with nutty flavour, Australia (Irvine 1957).

Lepidosperma squamatum *Labill.*, Nov. Holl. Pl. 1: 17, t. 16 (1805).
Australia; heaths, open areas.

MATERIALS: *Cane/reed*: matting, Australia (H290 — Fig. 2D).

LEPIRONIA *Rich.*

Lepironia articulata (*Retz.*) *Domin*, Biblioth. Bot. Heft. 85: 486 (1915).
Restio articulatus Retz., Observ. Bot. 4: 14 (1786).
Madagascar, S China, Sri Lanka, to Australia, Pacific Is.; open, swampy places and marshes.

FOOD: *Tubers*: edible, Australia (Irvine 1957).

MATERIALS: *Fibres*: dried culms, mats for packing tobacco, rubber and kapok, S Sumatra and Borneo (Kern 1974); culms and leaves used for packing, matting and basketwork, sometimes cultivated, China, Madagascar, New Guinea, Thailand (Heywood 1993; Kern 1974; D. A. Simpson, pers. comm. 1993; Simpson & Koyama 1998; H289 — Fig. 2B, H322); *Other materials/chemicals*: (Heywood 1993).

LIPOCARPHA *R. Br.*

Lipocarpha chinensis (*Osbeck*) *J. Kern*, Blumea Suppl. 4: 167 (1958).
Scirpus chinensis Osbeck, Dagb. Ostind. Resa: 220 (1757).
Widespread in the Old World tropics; open, wet areas.

ANIMAL FOOD: *Unspecified part*: plant provides a little grazing for livestock (Burkill 1985).

SOCIAL USES: '*Religious*' *uses*: some of plant is carried in hair or on wrists, magical properties to overcome crises in madness, Gabon (Burkill 1985).

MEDICINES: *Pain*: plant ash is rubbed with sap from a lime into scarifications on temple and forehead as an analgesic for headache, Tanzania (Burkill 1985).

WEEDS: *Grassland*: (Kühn 1982); *Waste places* (Kühn 1982); *Aquatic biotopes*: (Kühn 1982); *Rice fields*: (Kühn 1982; Simpson & Koyama 1998).

Lipocarpha kernii (*Raymond*) *Goetgh.*, Agric. Univ. Wageningen Pap. 89: 42 (1989).
Scirpus kernii Raymond, Naturaliste Canad. 86: 230 (1959).
Subsharan Africa south to Zimbabwe, India; wet, muddy and swampy areas.
ANIMAL FOOD: *Unspecified part*: cattle, Senegal (Burkill 1985).

Lipocarpha squarrosa (*L.*) *Goetgh.*, Agric. Univ. Wageningen Pap. 89: 71 (1989).
Scirpus squarrosus L., Mant. Pl. 2: 181 (1771).
India, Sri Lanka, Indo-China, Malayisa; open, wet, sandy or clay soils.
WEEDS: *Rotation crops*: soya bean fields, Thailand (Radanachaless & Maxwell 1994).

MACHAERINA *Vahl*

Machaerina angustifolia (*Gaudich.*) *T. Koyama*, Bot. Mag. (Tokyo) 69: 62 (1956).
Vincentia angustifolia Gaudich. in Freyc., Voy. Uranie: 417 (1829).
Hawaii; open boggy areas in forest and scrub.
MATERIALS: *Fibres*: leaves used for tying and lashing thatch to houses, Hawaii (Funk 1978).

Machaerina gunnii (*Hook. f.*) *J. Kern*, Acta Bot. Neerl. 8: 266 (1959).
Cladium gunnii Hook. f., Fl. Tasman. 2: 95 (1858).
Australia, New Guinea; swampy grassland.
MATERIALS: *Fibres*: culms, used to make skirts, New Guinea (Kern 1974).

Machaerina mariscoides (*Gaudich.*) *J. Kern*, Acta Bot. Neerl. 8: 266 (1959).
Baumea mariscoides Gaudich. in Freyc., Voy. Bot. 417 (1829).
New Guinea, Pacific Is.; secondary forest and open hillsides.
MATERIALS: *Fibres*: leaves used for tying and lashing thatch to houses (Funk 1978).

Machaerina rubiginosa (*Spreng.*) *T. Koyama*, J. Fac. Sci. Univ. Tokyo, Sect. 3, Bot. 8: 123 (1961).
Fuirena rubiginosa Spreng., Mant. Prim. Fl. Hal.: 29 (1807).
India, Sri Lanka to New Zealand, New Caledonia; swamps, lake margins.
MATERIALS: *Fibres*: leaves used to make mats, temporary tying material and to plait children's baskets, inferior quality, not damp-proof, New Guinea, Sumatra, W Java (Kern 1974); used for making bags, Australia (H281 — Fig. 1A).

Machaerina sinclairii (*Hook. f.*) *T. Koyama*, Bot. Mag. (Tokyo) 69: 65 (1956).
Cladium sinclairii Hook. f., Handb. N. Zeal. Fl. 1: 305 (1864).
Sumatra, Philippines, Sulawesi, Maluku, New Zealand; open hillsides and mountain slopes.
ENVIRONMENTAL USES: *Ornamentals*: (Grounds 1989).

MAPANIA *Aubl.*

Mapania bancana (*Miq.*) *Benth. & Hook. f. ex B. D. Jacks.*, Index Kew. 2: 163 (1895).
Lepironia bancana Miq., Fl. Ned. Ind., Eerste Bijv.: 263 (1862).
Thoracostachyum bancanum (Miq.) Kurz, J. Asiat. Soc. Bengal, Pt. 2, Nat. Hist. 38: 76 (1869).
Peninsular Thailand and Malaysia, Sumatra, Borneo, Sulawesi, New Guinea; peat swamp forest, flooded areas in kerangas forest.
MEDICINES: *Unspecified medicinal disorders*: Sabah (Simpson 1992b).

Mapania cuspidata (*Miq.*) *Uittien*, J. Arnold Arbor. 20: 213 (1939).
Lepironia cuspidata Miq., Fl. Ned. Ind., Eerste Bijv.: 263, 603 (1862).
Nicobar Is., through Malesia to the Solomon Is. and New Hebrides; evergreen forests.
MEDICINES: *Infections/infestations*: leaves used as a remedy against fever, Malay Peninsula (Burkill 1935; Kern 1974; Simpson 1992b).
ENVIRONMENTAL USES: *Ornamentals*: (Simpson 1992b).

Mapania kurzii *C. B. Clarke* in Hook. f., Fl. Brit. India 6: 681 (1894).
Malay Peninsula and Sumatra; evergreen forests.
ANIMAL FOOD: *Unspecified part*: fodder, Malaysia (Burkill 1935).

Mapania linderi *Hutch. ex Nelmes*, Kew Bull. 6: 422 (1951).
Western Africa (Guinea, Sierra Leone, Liberia, Ivory Coast); riverine forests, forests on white sand.
MATERIALS: *Fibres*: leaves reportedly used for roof-thatching, Liberia (Burkill 1985).

Mapania mannii *C. B. Clarke* in Th. Durand & Schinz, Consp. Fl. Afr. 5: 667 (1894); in Dyer, Fl. Trop. Afr. 8: 491 (1902).
West and central Africa; forest.
ENVIRONMENTAL USES: *Ornamentals*: (Simpson 1992b).

Mapania palustris (*Steud.*) *Fern.-Vill.*, Nov. App.: 309 (1882).
Pandanophyllum palustre Hassk. ex Steud., Syn. Pl. Glumac. 2: 134 (1855).
Malay Peninsula, Singapore, Sumatra, Java, Borneo, Philippines, Maluku, New Guinea, Solomon Is., New Hebrides; damp, shady areas in primary and secondary wet forest.
MATERIALS: *Fibres*: mats and baskets, Peninsular Malaysia (Burkill 1935), Brunei (Simpson 1992b).
ENVIRONMENTAL USES: *Ornamentals*: (Simpson 1992b).

Mapania sumatrana (*Miq.*) *Benth.*, Fl. Austral. 7: 341 (1878).
Lepironia sumatrana Miq., Fl. Ned. Ind., Eerste Bijv.: 263, 604 (1862).
Thoracostachyum sumatranum (Miq.) Kurz, J. Asiat. Soc. Bengal, Pt. 2, Nat. Hist. 38: 75 (1869).

Sumatra, Java, Borneo; swamps, marshy areas which are periodically flooded; margins of peat swamp forest.

MATERIALS: *Fibres*: leaves, mat-making, sometimes cultivated, Sumatra and Malay Peninsula (Burkill 1935; Kern 1974).

NEMUM *Desv.*

Nemum spadiceum (*Lam.*) *Desv. ex Ham.*, Prodr. Pl. Ind. Occid.: 13 (1825).
Eriocaulon spadiceum Lam., Tabl. Encycl. 2: 214 (1792).
Tropical Africa, W Indies; rocky areas, seepage areas.

ANIMAL FOOD: *Unspecified part*: provides a little grazing, Senegal (Burkill 1985).

OREOBOLUS *R. Br.*

Oreobolus pectinatus *Hook. f.*, Fl. Antarct. 1: 87, t. 49 (1847).
New Zealand; moist habitats.

ENVIRONMENTAL USES: *Ornamentals*: (Grounds 1989).

OXYCARYUM *Nees*

Oxycaryum cubensis (*Poepp. & Kunth*) *Lye*, Bot. Not. 124: 281(1971).
Scirpus cubensis Poepp. & Kunth in Kunth, Enum. Pl. 2: 172 (1837).
Tropical Africa, C and S America; floating in ponds, lakes and slow-moving rivers.

SOCIAL USES: '*Religious uses*': Shipibo-Conibo fisherman bathe in water containing the crushed rhizomes, and before fishing small pieces of rhizome are crushed on the prow of the canoe and rubbed on to the cord of the harpoon, Peru (Tournon *et al.* 1986).

WEEDS: *Aquatic biotopes*: Ghana (Abbiw 1990).

PSEUDOSCHOENUS *Oteng-Yeb.*

Pseudoschoenus inanis (*Thunb.*) *Oteng-Yeb.*, Notes Roy. Bot. Gard. Edinburgh 33 (2): 309 (1974).
Scirpus inanis Thunb., Prodr. Fl. Cap. 1: 16 (1794).
S Africa; wet places.

MATERIALS: *Fibres*: mats for building traditional beehive huts, S Africa (Van Eyk & Gericke 2000).

PYCREUS *P. Beauv.*

Pycreus elegantulus (*Steud.*) *C. B. Clarke* in T. Durand & Schinz, Consp. Fl. Afr. 5. 194: 536 (1895); in Dyer, Fl. Trop. Afr. 8: 302 (1902).
Cyperus elegantulus Steud., Flora 25: 583 (1842).
Subsaharan Africa; wet grassland swamps, stream margins

ANIMAL FOOD: *Unspecified part*: grazed by all domestic livestock, Kenya (Burkill 1985).

Pycreus filicinus (*Vahl*) *T. Koyama*, Phytologia 29 (2): 74 (1974).
Cyperus filicinus Vahl, Enum. Pl. 2: 332 (1806).
Eastern U.S.A., W Indies; margins of water bodies, dune depressions.
  ENVIRONMENTAL USES: *Ornamentals*: (Huxley 1992).

Pycreus flavidus (*Retz.*) *T. Koyama*, J. Jap. Bot. 51: 316 (1976).
Cyperus flavidus Retz., Observ. Bot. 5: 13 (1789).
Old World tropics and temperate regions; open wet places, rice fields, river margins, swamps, ponds.
  WEEDS: *Unspecified weed*: (USDA-ARS 2000).

Pycreus macrostachyos (*Lam.*) *J. Raynal*, Kew Bull. 23: 314 (1969).
Cyperus macrostachyos Lam., Tabl. Encycl. 1: 147 (1791).
Subsaharan Africa; seasonally wet areas, rock pools, wallows, damp grasslands.
  FOOD: *Rhizomes*: said to be eaten (Burkill 1985).

Pycreus muricatus (*Kük.*) *Napper*, J. E. Africa Nat. Hist. Soc. Natl. Mus. 28 (124): 6 (1971).
Cyperus muricatus Kük., Repert. Spec. Nov. Regni Veg. 12: 92 (1913).
E and southern Africa; boggy grasslands.
  FOOD: *Rhizomes*: Tanzania (Altschul 1973).
  MATERIALS: *Other materials/chemicals*: rhizome used for a perfume, Tanzania (Altschul 1973).

Pycreus mundtii *Nees*, Linnaea 10: 131 (1836).
Cyperus mundtii (Nees) Kunth, Enum. Pl. 2: 17 (1837).
Mediterranean, Africa, Mascarenes, W Indies; wet habitats, often floating.
  MATERIALS: *Unspecified material type*: culm said to be suitable for pulping but not for thatching (Burkill 1985).
  SOCIAL USES: '*Religious*' *uses*: whole plant is soaked in water, the water is then drunk and rubbed on the face for protection against very bad spirits, Madagascar; used in ceremony by fortune-teller to change a person's destiny, Madagascar (Beaujard 1988).
  MEDICINES: *Ill-defined symptoms*: medicines; may be smoked during sickness, Madagascar (Beaujard 1988).

Pycreus nitidus (*Lam.*) *J. Raynal*, Kew Bull. 23: 314 (1969).
Cyperus nitidus Lam., Tabl. Encycl. 1: 145 (1791).
Subsharan Africa; swamps and swamp margins.
  FOOD ADDITIVES: *Entire plant*: before commercial salt, plants were burnt and the ash lixivated to obtain a cooking salt, Uganda (Burkill 1985).
  MATERIALS: *Fibres*: culms used for making mats, Tanzania (Burkill 1985); used for matting, E Africa (Greenway 1950); *Other materials/chemicals*: scented rhizome used to freshen clothes, Lesotho (Burkill 1985).
  VERTEBRATE POISONS: *Mammals*: plant suspected of causing 'vlei' disease of sheep, S Africa (Burkill 1985).

MEDICINES: *Respiratory system disorders*: rhizome made into medicine for chest-coughs, Lesotho (Burkill 1985).

Pycreus polystachyos (*Rottb.*) *P. Beauv.*, Fl. Oware 2: 48, t. 86, f. 2 (1807).
Cyperus polystachyos Rottb., Descr. Icon. Rar. Pl.: 39, t. 11, f. 1 (1773).
Tropics, subtropics and warm temperate regions worldwide; open, moist or rather dry ground, frequently seen in sandy soil near seashores, grassy fields, waysides and on river banks.
   ANIMAL FOOD: *Unspecified part*: grazed by cattle, Tanzania, fed to livestock, Vietnam (Burkill 1985).
   WEEDS: *Grassland*: (Kühn 1982); *Waste places*: (Kühn 1982; Healy & Edgar 1980); *Aquatic biotopes*: (Kühn 1982); *Rice fields*: (Kühn 1982); *Pasture*: (Healy & Edgar 1980).

var. **laxiflorus** (*Benth.*) *C. B. Clarke* in Hook. f., Fl. Brit. India 6: 592 (1893).
Cyperus polystachyos Rottb. var. *laxiflorus* Benth., Fl. Austral. 7: 261 (1878).
SE Asia, Australia; similar to the above.
   WEEDS: *Rice fields*: (Burkill 1985).

Pycreus sanguinolentus (*Vahl*) *Nees*, Linnaea 9: 6 (1835).
Cyperus sanguinolentus Vahl, Enum. Pl. 2: 351 (1806).
Tropical Africa, India and Sri Lanka to W Malesia, China and Japan, Australia; wet grasslands, ditches, margin of swamps and rice fields.
   WEEDS: *Grassland*: (Kühn 1982); *Waste places*: (Healy & Edgar 1980); *Aquatic biotopes*: (Kühn 1982); *Rice fields*: (Kühn 1982); *Pasture*: (Healy & Edgar 1980).

QUEENSLANDIELLA *Domin*

Queenslandiella hyalina (*Vahl*) *Ballard*, Hooker's Icon. Pl.: 33, t. 3208 (1933).
Cyperus hyalinus Vahl, Enum. Pl. 2: 329 (1806).
Tropical E Africa to northern Australia; sandy soils, usually coastal.
   WEEDS: *Disturbed land*: weed of disturbed ground of fishing village outside city, Tanzania (H95); *Gardens*: weed in lawns, Kenya (H94), weed of rough cut turf in gardens, Tanzania (H96).

REMIREA *Aubl.*

Remirea maritima *Aubl.*, Hist. Pl. Guiane 1: 45, t. 16 (1775).
Pantropical; coastal sand-dunes and seashores.
   ANIMAL FOOD: *Leaves/culms/aerial parts*: goats browse leaves (Burkill 1985).
   MATERIALS: *Other materials/chemicals*: rhizome aromatic, Brazil (Pio Corrêa 1926).
   MEDICINES: *Unspecified medicinal disorders*: rhizome astringent, infusion said to be used as a sudorific, diuretic, diaphoretic and anti-blenorrhagic, Brazil, Guyana (Burkill 1985; Cauis & Banby 1935; Pio Corrêa 1926); often burnt to release odour, Brazil (Pio Corrêa 1926); *Genito-urinary system disorders*: infusion of rhizome said to be used as a diuretic, Brazil, Guyana (Burkill 1985; Cauis & Banby 1935).

ENVIRONMENTAL USES: *Erosion control*: sand-binder (Anon. 1994 – 2000), Malay Peninsula, Africa (Burkill 1935; Burkill 1985).

RHYNCHOSPORA *Vahl*

Rhynchospora barbata (*Vahl*) *Kunth*, Enum. Pl. 2: 290 (1837).
Schoenus barbatus Vahl, Eclog. Amer. 2: 4 (1798).
Central and tropical S America; savannahs.
  ANIMAL FOOD: *Unspecified part*: forage, Brazil (Pio Corrêa 1926).

Rhynchospora cephalotes (*L.*) *Vahl*, Enum. Pl. 2: 237 (1805).
Scirpus cephalotes L., Sp. Pl. ed. 2: 76 (1762).
Tropical America; savannahs, open river banks.
  ANIMAL FOOD: *Unspecified part*: forage, Brazil (Pio Corrêa 1926).
  MATERIALS: *Fibres*: used to make a translucent paper, Brazil (Pio Corrêa 1926); used for weaving mats, bottle covers, chairs etc, Brazil (Pio Corrêa 1926).

Rhynchospora colorata (*L.*) *Pfeiff.*, Repert. Spec. Nov. Regni. Veg. 38: 89 (1935).
Schoenus coloratus L., Sp. Pl. 1: 43 (1753).
Dichromena colorata (L.) Hitchc., Annual Rep. Missouri Bot. Gard. 4: 141 (1893).
Southern U.S.A., C and northern S America; open sandy soils, roadsides, pastures.
  ENVIRONMENTAL USES: *Ornamentals*: (Lord *et al.* 2000).

Rhynchospora corymbosa (*L.*) *Britton*, Trans. New York Acad. Sci. 11: 84 (1892).
Scirpus corymbosus L., Cent. Pl. 2: 7 (1756).
Pantropical; open, swampy places.
  MATERIALS: *Fibres*: rough string, Tanzania (Burkill 1985); mats, sandals, screens, baskets, of little value, Philippines (Burkill 1985; Kern 1974); used for string, E Africa (Greenway 1950).
  SOCIAL USES: '*Religious*' *uses*: had ceremonial use, sign of messenger/herald, Bataks of Sumatra (Burkill 1935).
  MEDICINES: *Digestive system disorders*: decoction of nutlet relieves children's abdominal pains and colic, Gambia, Nigeria (Burkill 1985).
  ENVIRONMENTAL USES: *Revegetators*: colonises badly drained land, put to use in tsetse fly control, W Africa (Burkill 1985); *Soil improvers*: ploughed in as green manure, SE Asia (Burkill 1935), W Africa (Burkill 1985).
  WEEDS: *Waste places* (Kühn 1982); *Aquatic biotopes*: (Kühn 1982); *Rice fields*: Sierra Leone (Burkill 1985).

Rhynchospora hirsuta (*Vahl*) *Vahl*, Enum. Pl. 2: 231 (1805).
Schoenus hirsutus Vahl, Eclog. Amer. 1: 6 (1796).
Central America, W Indies, tropical S America; savannahs.
  ANIMAL FOOD: *Unspecified part*: forage, Brazil (Pio Corrêa 1926).

Rhynchospora holoschoenoides (*Rich.*) *Herter*, Revista Sudamer. Bot. 9: 157 (1953).
Schoenus holoschoenoides Rich., Actes Soc. Hist. Nat. Paris 1: 106 (1792).

Tropical Africa and America; wet areas near swamps or streams.
ANIMAL FOOD: *Unspecified part*: forage, Brazil (Pio Corrêa 1926).
WEEDS: *Rice fields*: Sierra Leone (Burkill 1985).

Rhynchospora nervosa (*Vahl*) *Boeck.*, Vidensk. Meddel Dansk Naturhist. Foren. Kjøbenhavn, ser. 3, 1: 143 (1869).
Dichromena nervosa Vahl, Enum. Pl. 2: 241 (1806).
Greater Antilles, C America to Argentina; savannahs.
MEDICINES: *Infections/infestations*: infusion of whole plant used by Tiriyó as external bath for fevers, Brazil (Milliken 1997).
ENVIRONMENTAL USES: *Ornamentals*: (Huxley 1992; Simpson 1993).

subsp. **ciliata** (*Vahl*) *T. Koyama*, Madroño 20: 254 (1970).
Dichromena ciliata Vahl, Enum. Pl. 2: 240 (1806).
Southern Mexico and the Antilles to Bolivia and southern Brazil; weedy areas, particularly pastures, roadsides and lawns.
ANIMAL FOOD: *Unspecified part*: forage, Brazil (Pio Corrêa 1926).
MATERIALS: *Fibres*: filling for mattresses, Dominican Republic (Simpson 1993); *Other materials/chemicals*: rhizome aromatic (Pittier 1971).
MEDICINES: *Unspecified medicinal disorders*: used in popular medicine, Venezuela (Pittier 1971).
ENVIRONMENTAL USES: *Ornamentals*: (Huxley 1992).
WEEDS: *Unspecified weed* (Simpson 1993); *Rotation crops*: (Kühn 1982); *Perennial crops* (Kühn 1982); *Grassland*: (Kühn 1982); *Aquatic biotopes*: (Kühn 1982).

Rhynchospora rubra (*Lour.*) *Makino*, Bot. Mag. (Tokyo) 17: 180 (1903).
Schoenus ruber Lour., Fl. Cochinch. 1: 41 (1790).
India to Vietnam (type), S China, central Japan, Malesia to Australasia and Pacific Is.; open grasslands and hillsides, sometimes grassy waysides.
MEDICINES: *Unspecified medicinal disorders:* whole plant used to relieve fever, China (J. C. Shaw, C. S. Tong & L. Wong pers. comm. 1995).

Rhynchospora tenerrima *Nees ex Spreng.*, Syst. Veg. 4: 26 (1827).
Tropical America; savannahs and moist open areas.
ANIMAL FOOD: *Unspecified part*: forage, Brazil (Pio Corrêa 1926).

SCHOENOPLECTUS (*Rchb.*) *Palla*

Schoenoplectus americanus (*Pers.*) *Volkart* in Schinz & R. Keller, Fl. Schweiz, ed. 2: 75 (1905).
Scirpus americanus Pers., Syn. Pl. 1: 68 (1805).
Europe, S America, S Australia; salt marshes, mudflats.
FOOD: *Unspecified part*: vegetable (Anon. 1994 – 2000).
MATERIALS: *Fibres*: culm (Anon. 1994 – 2000).

Schoenoplectus articulatus (*L.*) *Palla*, Bot. Jahrb. Syst. 10: 299 (1888).

Scirpus articulatus L., Sp. Pl. 1: 47 (1753).

Mediterranean region and Africa through India and Sri Lanka to SE China, Malesia and N Australia; open marshy places.

MEDICINES: *Digestive system disorders*: used as a purgative, India (Cauis & Banby 1935; Mukhopadhyay & Ghosh 1992); rhizomes, Hindu medicine, purgative (Heywood 1993).

WEEDS: *Rice fields*: (Simpson & Koyama 1998).

Schoenoplectus californicus (*C. A. Mey.*) *Soják*, Čas. Nár. Mus., Odd. Přír. 140 (3 – 4): 127 (1972).

Elytrospermum californicum C. A. Mey., Mém. Acad. Imp. Sci. St.-Pétersbourg Divers Savans 1: 201 (1831).

Scirpus californicus (C. A. Mey.) Steud., Nomencl. Bot. 2, 2: 538 (1841).

Southern U.S.A., C and S America; wet places.

ANIMAL FOOD: *Unspecified part*: medium potential for grazing, U.S.A. (USDA-NRCS 1999).

MATERIALS: *Fibres*: Mexico, Peru (USDA-ARS 2000); matting, Ecuador (H287 — Fig. 1H).

ENVIRONMENTAL USES: *Erosion control*: U.S.A. (USDA-ARS 2000).

Schoenoplectus corymbosus (*Roem. & Schult.*) *J. Raynal*, Cat. Pl. Vasc. Niger, Cyperac.: 343 (1976).

Isolepis corymbosa Roth ex Roem. & Schult., Syst. Veg. 2: 110 (1817).

Africa to India; open water up to 1 m deep.

MATERIALS: *Fibres*: culms used for making sleeping mats, E Africa (Greenway 1950).

VERTEBRATE POISONS: *Mammals*: poisonous to cattle, S Africa (Cauis & Banby 1935).

var. **brachyceras** (*A. Rich.*) *Lye*, Nordic J. Bot. 3 (2): 242 (1983).

Scirpus brachyceras Hochst. ex A. Rich., Tent. Fl. Abyss. 2: 496 (1851).

Schoenoplectus brachyceras (A. Rich.) Lye, Bot. Not. 124 (2): 290 (1971).

Africa; bogs and margins of water bodies

FOOD: *Culms*: Kipsigi children chew the white culm tips, Kenya (Burkill 1985).

ANIMAL FOOD: *Leaves/culms/aerial parts*: domestic livestock graze culms, Kenya (Burkill 1985).

MATERIALS: *Fibres*: Ancient Egyptians used culms for making funeral wreaths, bound culms were ornated with flowers, berries etc., sometimes peeled culms were used for making artificial flowers (Täckholm & Drar 1950); used to make traditional zulu sleeping mats, S Africa (Van Eyk & Gericke 2000).

Schoenoplectus dissachanthus (*S. T. Blake*) *J. Raynal*, Adansonia sér. 2, 16: 139 (1976).

Scirpus dissachanthus S. T. Blake, Victoria Naturalist 63: 116 (1946).

Australia; muddy areas on swamp margins, grasslands, scrub.

FOOD: *Nutlets*: (Anon. 1994 – 2000).

Schoenoplectus junceus (*Willd.*) *J. Raynal*, Adansonia sér. 2, 16: 139 (1976).
Schoenus junceus Willd., Phytographia 1: 2, t. 1, 4 (1794).
Tropical Africa; seasonally wet areas, particularly with standing water.

MEDICINES: *Poisonings*: pounded in water and applied externally as antidote to scorpion-sting, W Africa (Abbiw 1990; Burkill 1985).

Schoenoplectus juncoides (*Roxb.*) *Palla*, Bot. Jahrb. Syst. 10: 229 (1889).
Scirpus juncoides Roxb., Fl. Ind. 1: 218 (1820).
Madagascar, India to N Australia; open, wet localities.

ANIMAL FOOD: *Unspecified part*: cattle, feeding value high (Kern 1974).

MEDICINES: *Unspecified medical disorders*: used to release heat, China (J. C. Shaw, C. S. Tong & L. Wong pers. comm. 1995); *Respiratory system disorders*: used to stop coughing, China (J. C. Shaw, C. S. Tong & L. Wong pers. comm. 1995); *Sensory system*: used to clear eyes, China (J. C. Shaw, C. S. Tong & L. Wong pers. comm. 1995).

WEEDS: *Rotation crops*: soya bean fields, Thailand (Radanachaless & Maxwell 1994); *Aquatic biotopes*: (Kühn 1982); *Rice fields*: (Kühn 1982; Simpson & Koyama 1998).

Schoenoplectus lacustris (*L.*) *Palla*, Verh. K. K. Zool.-Bot. Ges. Wien 38: 49 (1888).
Scirpus lacustris L., Sp. Pl. 1: 48 (1753).
Widespread in temperate to tropical regions; swampy areas, margins of water bodies.

FOOD: *Unspecified part*: vegetable (Anon. 1994 – 2000).

MATERIALS: *Fibres*: culm (Anon. 1994 – 2000); culms used for basketwork, protective sleeve cuffs (Heywood 1993; H292 — Fig. 2F); chair seating, U.K. (Lewington 1990); still used in some parts of the world to make matting (Heywood 1993; Lewington 1990; H294 — Fig. 2H).

MEDICINES: *Unspecified medicinal disorders*: culms (Anon. 1994 – 2000); astringent, Europe (Cauis & Banby 1935); *Genito-urinary system disorders*: diuretic, Europe (Cauis & Banby 1935).

WEEDS: *Aquatic biotopes*: (Kühn 1982).

subsp. **validus** (*Vahl*) *T. Koyama*, Bishop Mus. Occas. Pap. 29: 128 (1989).
Scirpus validus Vahl, Enum. Pl. 2: 268 (1806).
Countries bordering the Pacific Ocean; wet or flooded soils.

FOOD: *Rhizomes*: cooked vegetable (Anon. 1994 – 2000).

MATERIALS: *Fibres*: weaving mats, N Luzon, Philippines (Kern 1974); culms uesd for mats forming lower layers of sleeping couches (Funk 1978); temporary mats were made for wrapping bodies, material was not durable (Funk 1978); culms used for thatching (Funk 1978).

ENVIRONMENTAL USES: *Ornamentals*: N America (Huxley 1992); *Pollution control*: used for cleaning up water in lagoons, U.S.A. (Anon. 1990).

Schoenoplectus lateriflorus (*J. F. Gmel.*) *Lye*, Bot. Not. 124 (2): 290 (1971).
Scirpus lateriflorus J. F. Gmel., Syst. Nat. 1: 127 (1791).
India to Australia; open wet places.

ANIMAL FOOD: *Unspecified part*: fodder, cattle (Burkill 1935).

MATERIALS: *Fibres*: used for making hats, Taiwan (H291 — Fig. 2E).

ENVIRONMENTAL USES: *Soil improvers*: ploughed in as green manure (Burkill 1935).

This plant is sometimes referred to *Schoenoplectus erectus* (Poir.) J. Raynal which, although closely related, occurs in Africa and the Mascarenes.

Schoenoplectus littoralis (*Schrad.*) *Palla*, Verh. K. K. Zool.-Bot. Ges. Wien 38: 49 (1888).

Scirpus subulatus Vahl, Enum. Pl. 2: 268 (1806).

Scirpus litoralis Schrad., Fl. Germ. 1: 42. t. 5. f. 7 (1806).

Schoenoplectus subulatus (Vahl) Lye, Bot. Not. 124: 290 (1971).

Schoenoplectus littoralis (Schrad.) Palla subsp. *subulatus* (Vahl) T. Koyama in Dassan. & Fosberg, Rev. Handb. Fl. Ceylon 5: 157 (1985).

Mediterranean region and Africa to Australia; brackish, swampy places, often coastal.

FOOD: *Rhizomes*: cooked vegetable (Anon. 1994 – 2000); *Culms*: roasted in sections then peeled and eaten, or the outer skin of culm removed to reveal white interior, eaten raw or boiled, Oman (Millar & Morris 1988).

ANIMAL FOOD: *Leaves/culms/aerial parts*: tender tops eaten by certain water-birds, eaten by cattle only when nothing else is available, Egypt (Täckholm & Drar 1950); leaves fed to livestock in times of scarcity of other fodder, Oman (Millar & Morris 1988).

MATERIALS: *Unspecified material type*: culms have been used in the past for caulking palm-oil casks (Burkill 1985); *Fibres*: culms cut to stuff mattresses, Ghana (Burkill 1985); used to make mats, W Java (Kern 1974); body of the plant makes good roofing material, Oman (Millar & Morris 1988).

WEEDS: *Aquatic biotopes*: (Kühn 1982).

Schoenoplectus mucronatus (*L.*) *Palla*, Verh. K. K. Zool.-Bot. Ges. Wien 38: 49 (1888).

Scirpus mucronatus L., Sp. Pl. 50. 1753.

Southern Europe to Japan, Asia and Australia, rare in Africa; open wet places.

MATERIALS: *Fibres*: matting, commonly used as string, Malay Peninsula (Burkill 1935); culms woven into cheap and durable mats, bags and string, cultivated, Sumatra (Kern 1974).

WEEDS: *Aquatic biotopes*: (Kühn 1982); *Rice fields*: (Kühn 1982; Simpson & Koyama 1998).

Schoenoplectus muricinux (*C. B. Clarke*) *J. Raynal*, Adansonia sér. 2, 15: 538 (1976).

Scirpus muricinux C. B. Clarke, Bot. Jahrb. Syst. 38: 135 (1906).

Southern Africa; seasonally wet grasslands.

ANIMAL FOOD: *Unspecified part*: eaten by cattle and goats, Namibia (Rodin 1985).

Schoenoplectus paludicola (*Kunth*) *Palla ex J. Raynal*, Adansonia sér. 2, 15: 541 (1976).

Scirpus paludicola Kunth, Enum. Pl. 2: 163 (1837).
S Africa; margins of pools and streams.
   MEDICINES: *Unspecified medicinal disorders*: S Africa (Cauis & Banby 1935).

Schoenoplectus pungens (*Vahl*) *Palla*, Bot. Jahrb. Syst. 10: 299 (1888).
Scirpus pungens Vahl, Enum. Pl. 2: 255 (1806).
W Europe, America, Australia, New Zealand; salt marshes, brackish swamps, usually coastal sometimes inland.
   WEEDS: *Unspecified weed*: troublesome in coastal areas, New Zealand (Healy & Edgar 1980).

Schoenoplectus riparius (*C. Presl*) *Palla*, Bot. Jahrb. Syst. 10: 299 (1888).
Scirpus riparius C. Presl in J. Presl & C. Presl, Reliq. Haenk. 1: 193 (1828).
N, C and S America, Hawaii; standing water or mud on margins of water bodies.
   MATERIALS: *Cane/reed*: boats, cut, dried and lashed together in two thick bundles about 3.5 m long, tied side-on to form the raft-like buoyant base, with the rear end blunt and front end curving to a sharply tapered point. Fishermen kneel or sit astride the 'caballitos de totora' and use a paddle to manoeuvre them, plants growing in the vicinity used, Peru (Lewington 1990).

Schoenoplectus roylei (*Nees*) *Ovcz. & Czukav.*, Fl. Tadjikist. 2: 40 (1963).
Isolepis roylei Nees in Wight, Contr. Bot. India: 107 (1834).
Africa to India; seasonally wet grassland.
   WEEDS: *Rice fields*: Kenya (Burkill 1985); *Irrigation ditches*: Kenya (Burkill 1985).

Schoenoplectus scirpoideus (*Schrad.*) *J. Browning*, S. African J. Bot. 60(3): 172 (1994).
Pterolepis scirpoides Schrad., Gött. Gel. Anz. 3: 2071 (1821).
S Africa; estuarine localities, coastal pools.
   MATERIALS: *Fibres*: sleeping mats, S Africa (Van Eyk & Gericke 2000).

Schoenoplectus senegalensis (*Hochst. ex Steud.*) *Palla ex J. Raynal*, Cat. Pl. Vasc. Niger, Cyperac.: 344 (1976).
Isolepis senegalensis Hochst. ex Steud., Syn. Pl. Glumac. 2: 96 (1855).
Tropical Africa; seasonally wet areas and margins of permanent pools.
   NON-VERTEBRATE POISONS: *Mollusca*: used as molluscicide, Sudan (Ahmed *et al.* 1984).

Schoenoplectus supinus (*L.*) *Palla*, Bot. Jahrb. Syst. 10: 299 (1888).
Scirpus supinus L., Sp. Pl. 1: 49 (1753).
Old World tropics and temperate regions; wet places.
   WEEDS: *Unspecified weed*: (USDA-ARS 2000); *Rice fields*: (Simpson & Koyama 1998).

Schoenoplectus tabernaemontani (*C. C. Gmel.*) *Palla*, Verh. K. K. Zool.-Bot. Ges. Wien. 38: 49 (1888).

Scirpus tabernaemontani C. C. Gmel., Fl. Bad. 1: 101 (1806).
Europe, temperate Asia; streams, ditches, pools and bogs.
    ENVIRONMENTAL USES: *Ornamentals*: 'Zebrinus' a distinctive ornamental plant (Grounds 1989; Huxley 1992).

Schoenoplectus tatora (*Kunth*) *Palla*, Bot. Jahrb. Syst. 10: 299 (1888).
Scirpus tatora Kunth, Enum. Pl. 2: 166 (1837).
Peru; lake margins.
    MATERIALS: *Cane/reed*: culms used to construct canoes and rafts, Uros and Aymar Indians, Lake Titicaca, Peru (Heywood 1993; Lewington 1990).

Schoenoplectus triqueter (*L.*) *Palla*, Verh. K. K. Zool.-Bot. Ges. Wien 38: 49 (1888).
Scirpus triqueter L., Mant. Pl.: 29 (1767).
Europe, temperate Asia, N Africa, N America; river banks.
    MATERIALS: *Fibres*: used for matting, China (H304); cultivated for 'taiko' mats, Taiwan (H324).
    WEEDS: *Aquatic biotopes*: (Kühn 1982).

Schoenoplectus wallichii (*Nees*) *T. Koyama*, Fl. Taiwan 5: 210 (1978).
Scirpus wallichii Nees in Wight, Contr. Bot. India: 112 (1834).
India to Japan and Malesia; wet places.
    MEDICINES: *Genito-urinary system disorders*: used in relief of cystitis, China (J. C. Shaw, C. S. Tong & L. Wong pers. comm. 1995).

SCHOENUS *L.*

Schoenus melanostachys *R. Br.*, Prodr.: 231 (1810).
Philippines, Borneo, Australia; open areas.
    MATERIALS: *Cane/reed*: used for basket making, Australia (H282 — Fig. 1B).

Schoenus nigricans *L.*, Sp. Pl. 1: 43 (1753).
Eurasia, S Africa, America; wet peaty places.
    MATERIALS: *Fibres*: used to make rope, Italy (H288 — Fig. 2A).

Schoenus pauciflorus (*Hook. f.*) *Hook. f.*, Handb. N. Zeal. Fl. 1: 298 (1864).
Chaetospora pauciflora, Hook. f., Fl. Nov.-Zel.: 273 (1853).
New Zealand; wet places in mountains.
    ENVIRONMENTAL USES: *Ornamentals*: (Grounds 1989; Huxley 1992).

SCIRPODENDRON *Zipp. ex Kurz*

Scirpodendron ghaeri (*Gaertn.*) *Merr.*, Philipp. J. Sci. 9: 268 (1914).
Chionanthus ghaeri Gaertn., Fruct. 1: 190, t. 39, f. 6a – e (1788).
Sri Lanka to N Australia, Pacific Is.; freshwater tidal areas, swamp forests.
    FOOD: *Nutlets*: Samoa (Kern 1974).

MATERIALS: *Fibres*: cultivated for mat making, S Sumatra, used to make hats, Philippines (Burkill 1935); dried leaves used for mats and hats, Sumatra, Leyte and Maluku (Kern 1974); dried leaves used for thatching, Fiji (Kern 1974); leaves used to weave mats Vanuatu (H173).

SCIRPOIDES *Ség.*

Scirpoides dioecus (Kunth) J. Browning, S. African J. Bot. 60 (6): 318 (1994).
Isolepis dioeca Kunth, Enum. Pl. 2: 199 (1837).
Scirpus dioecus Boeck., Linnaea 36: 719 (1869 – 70).
S Africa; open, wet, sandy or gravelly places.
   MATERIALS: *Fibres*: mats for building traditional beehive huts, S Africa (Van Eyk & Gericke 2000).

Scirpoides holoschoenus (*L.*) *Soják*, Čas. Nár. Mus., Odd. Přír. 140 (3 – 4): 127 (1972).
Scirpus holoschoenus L., Sp. Pl. 1: 49 (1753).
Europe, N Africa; damp sandy areas, coastal.
   MATERIALS: *Fibres*: used as a tying material, Canary Is. (H325).
   ENVIRONMENTAL USES: *Ornamentals*: including 'variegatus' (Huxley 1992).
   WEEDS: *Rotation crops*: (Kühn 1982); *Perennial crops* (Kühn 1982); *Waste places* (Kühn 1982).

subsp. **australis** (*Murray*) *Soják*, Čas. Nár. Mus., Odd. Přír. 141 (1 – 2): 61 (1972).
Scirpus australis Murray, Syst. Veg. 13: 85 (1774).
S Europe, N Africa; damp, sandy areas.
   MATERIALS: *Fibres*: culms used for matting, Libya (Anon. 1994 – 2000; Burkill 1985).

SCIRPUS *L.*

Scirpus cyperinus (*L.*) *Kunth*, Enum. Pl. 2: 170 (1837).
Eriophorum cyperinum L., Sp. Pl. ed 2, 1: 77 (1762).
N America; marshy meadows, swamp margins.
   ENVIRONMENTAL USES: *Ornamentals*: (Huxley 1992).

Scirpus sylvaticus *L.*, Sp. Pl. 1: 49 (1753).
Europe to Siberia, N America; marshes, wet places in woods, streamsides.
   ENVIRONMENTAL USES: *Ornamentals*: (Huxley 1992).

SCLERIA *Bergius*

Scleria biflora *Roxb.*, Fl. Ind. 2, 3: 573 (1832).
India and Sri Lanka to Sulawesi; grassy roadsides, waste places.
   FOOD: *Entire plant*: very young fragrant plants are eaten with rice, raw or steamed, Java (Kern 1974).

Scleria boivinii *Steud.*, Syn. Pl. Glumac. 2: 173 (1855).
Tropical Africa, Madagascar; swamp forest, forest margins.

MATERIALS: *Fibres*: matting, Ghana (Abbiw 1990); *Other materials/chemicals*: sharp-edged leaves used as 'razors', Congo (Burkill 1985); the leaves together with the leaves of *Staudtia stipitata* Warb. (*Myristicaceae*) used for catching, naked-handed, crustaceans and shellfish (Burkill 1985); nutlets used for beads and necklaces, Ghana (Abbiw 1990).

SOCIAL USES: '*Religious*' *uses*: young shoots with other herbs are cooked and given to secret society initiates to eat when the intoxicating effects of taking *Tabernanthe iboga* Baill. (*Apocynaceae*) are wearing off, Gabon (Burkill 1985); deemed to be a good talisman when the backs of hands are beaten (Burkill 1985); if scabrid parts are eaten they cause damage to the gut, criminal use in this way, Liberia, Ivory Coast (Burkill 1985); *Miscellaneous social uses*: whole plant used for aphrodisiac properties, Congo (Burkill 1985).

MEDICINES: *Unspecified medicinal disorders*: used as a mouthwash, Ivory Coast (Burkill 1985); *Genito-urinary system disorders*: rhizome-decoction to treat irregular menses or too abundant menstruation and also for haematuria, Congo (Burkill 1985); *Infections/infestations*: the dried, powdered rhizomes are applied topically over epidermal scarifications for leprous sores, Congo (Burkill 1985); rhizome decoction used against blennorrhoea, Gabon (Burkill 1985); whole plant used against blennorrhoea, Congo (Burkill 1985); *Pain*: a warm decoction is said to ease toothache, Ivory Coast (Burkill 1985); the dried powdered rhizomes are applied topically over epidermal scarifications for headaches, Congo (Burkill 1985); warm leaf decoction with *Scleria naumanniana* Boeck. and *S. depressa* (C. B. Clarke) Nelmes used as a mouthwash to relieve toothache, Ghana (Abbiw 1990); *Poisonings*: a leaf decoction from which all scabrid material is removed or avoided is used as a wash on snake and other animal bites, Ivory Coast (Burkill 1985); *Pregnancy/birth/puerperium disorders*: a leaf macerate is said to ease or hasten childbirth, Gabon (Burkill 1985); *Respiratory system disorders*: a leaf decoction from which all scabrid material is removed or avoided is taken for coughs, Ivory Coast (Burkill 1985); a rhizome macerate is taken in a draught for hiccups, Ivory Coast, Burkina Faso (Burkill 1985); the aerial parts are compounded into a formulation with several other plants and the liquid after cooking is taken as a drink for a cough, Gabon (Burkill 1985); *Sensory system*: aerial parts are used to relieve white patches of the cornea, culms, leaves and inflorescences are crushed with salt and the expressed liquid is administered as an eye-instillation, Ivory Coast (Burkill 1985).

Scleria depressa (*C. B. Clarke*) *Nelmes*, Amer. J. Bot. 39: 392 (1952).
Scleria racemosa Poir. var. *depressa* C. B. Clarke in Fl. Trop. Afr. 8: 508 (1902).
West and central tropical Africa; swamps in savannah and forest.

MATERIALS: *Fibres*: used for matting, Ghana; useful as thatch, Ghana (Abbiw 1990); *Other materials/chemicals*: blacksmiths burn culms in braziers to increase heat of charcoal, Uganda (Burkill 1985); nutlets used as beads, Ghana (Burkill 1985, Abbiw 1990).

MEDICINES: *Unspecified medicinal disorders*: the Ijo use leaves to make small cuts on the body into which to rub ointment, eg. over breasts for the administration of alligator pepper as a stimulant, Nigeria (Burkill 1985); *Genito-urinary system disorders*: rhizome-decoction taken with that of *Mimosa pigra* L. (*Leguminosae*) for dysmenorrhoea, Tanzania (Burkill 1985); *Infections/infestations*: rhizome decoction is taken for blennorrhoea, E Cameroun (Burkill 1985); a macerate of the pounded rhizome, for gonorrhoea, Tanzania (Burkill 1985); *Injuries*: sap from the base of the mucilaginous young shoots is applied to heal cuts and wounds, E Cameroun (Burkill 1985); *Pain*: warm leaf decoction with *S. boivinii* and *S. naumanniana* used as a mouthwash to relieve toothache, Ghana (Abbiw 1990).

Scleria foliosa *Hochst. ex A. Rich.*, Tent. Fl. Abyss. 2: 509 (1851).
Subsaharan Africa, Madagascar, India; wet grasslands, swamp and pond margins, cultivated areas.
    ANIMAL FOOD: *Unspecified part*: grazed by all livestock, Abyei district of S Kordofan, Sudan (Sudan Ministry 1980).
    MEDICINES: *Genito-urinary system disorders*: decoction of rhizomes is taken with *Mimosa pigra* for dysmenorrhoea, Tanzania (Burkill 1985); *Infections/infestations*: rhizome pounded in water and the macerate is taken for gonorrhoea, Tanzania (Burkill 1985).

Scleria hirtella *Sw.*, Prodr.: 19 (1788).
Tropical America, W Indies, tropical Africa; wet savannahs, bogs.
    MEDICINES: *Unspecified medicinal disorders*: rhizome oleo-aromatic, used in popular medicine, Venezuela (Pittier 1971).

Scleria iostephana *Nelmes*, Kew Bull. 11: 94 (1956).
Tropical Africa; forest or forest margins near swamps or rivers.
    MEDICINES: *Circulatory system disorders*: water in which rhizomes have been pulped is taken for haemorrhoids, Congo (Burkill 1985).

Scleria levis *Retz.*, Observ. Bot. 4: 13 (1786).
India and Sri Lanka to S China, Japan and N Australia, New Caledonia, Pacific Is.; open forests, savannahs.
    MEDICINES: *Respiratory system disorders*: nutlets good for coughs and are eaten with betel nut (Burkill 1935).

Scleria lithosperma (*L.*) *Sw.*, Prodr.: 18 (1788).
Scirpus lithospermus L., Sp. Pl. 1: 51 (1753).
Pantropical; semi-dry open grassy places, floors and margins of forests.
    ANIMAL FOOD: *Unspecified part*: fodder for goats, Nicobar Is. (Dagar & Dagar 1999).
    MEDICINES: *Genito-urinary system disorders*: decoction of plant taken in draught for dysmenhorrhoea, Tanzania (Burkill 1985); *Pregnancy/birth/puerperium disorders*: decoction of plant taken in a draught to stave off a threatened miscarriage, Tanzania (Burkill 1985); decoction of rhizomes may be administered after childbirth, Malaysia (Burkill 1935); *Respiratory system disorders*: entire plant pounded and squeezed into a bowl and the water of a young coconut is added before the mixture is drunk for the treatment of coughing (Altschul 1973).

WEEDS: *Perennial crops* (Kühn 1982); *Waste places* (Kühn 1982); *Aquatic biotopes*: (Kühn 1982); *Rice fields*: (Kühn 1982).

Scleria melanomphala *Kunth*, Enum. Pl. 2: 345 (1837).
Subsaharan Africa; open wet places.
   MEDICINES: *Genito-urinary system disorders*: whole plant including rhizome is decocted and the liquor is drunk for dysmenorrhoea, Tanzania (Burkill 1985).

Scleria mitis *Bergius*, Kongl. Vetensk. Acad. Handl. 27: 145, t. 5 (1765).
C America, W Indies, tropical S America; along streams in evergreen forests, swamps.
   ANIMAL FOOD: *Unspecified part*: eaten by cattle and horses, Venezuela (Altschul 1973).

Scleria naumanniana *Boeck.*, Bot. Jahrb. Syst. 5: 94 (1883).
W and central tropical Africa; damp places in forest or scrub, marshes, sand-dunes, mangrove swamps.
   MATERIALS: *Fibres*: used for matting, Ghana (Abbiw 1990); *Other materials/chemicals*: nutlets used as beads (Burkill 1985); nutlets useful for necklaces, Ghana (Abbiw 1990).
   MEDICINES: *Pain*: warm decoction of the leaves used as mouthwash to relieve toothache (Burkill 1985); warm decoction of leaves with *S. boivinii* and *S. depressa* used as a mouthwash to relieve toothache, Ghana (Abbiw 1990).

Scleria pergracilis (*Nees*) *Kunth*, Enum. Pl. 2: 354 (1837).
Hypoporum pergracile Nees, Edinburgh New Philos. J. 34: 267 (1834).
Tropical Africa to New Guinea; open slopes, swamp edges, savannahs.
   FOOD: *Leaves*: leaves eaten with salt, New Guinea (Kern 1974).
   MEDICINES: *Unspecified medicinal disorders*: used after confinement, Sumatra (Burkill 1935); *Infections/infestations*: lemon-scented leaves, used for fever, foot and mouth disease, Sumatra (Burkill 1935; Kern 1974); *Respiratory system disorders*: used in a decoction for coughs, Sumatra (Burkill 1935).

Scleria poaeformis *Retz.*, Observ. Bot. 4: 13 (1786).
Tropical E Africa to N Australia; freshwater swamps, swampy savannah forests, rice fields, ditches.
   MATERIALS: *Fibres*: leaves sometimes used for making mats, W Java (Kern 1974); *Other materials/chemicals*: rough leaves used for polishing wood (Burkill 1935).
   MEDICINES: *Inflammation*: fruiting panicles boiled and used for poulticing abdomen (Burkill 1935).

Scleria polycarpa *Boeck.*, Linnaea 28: 509 (1874).
New Guinea to N Australia, Pacific Is.; rainforests, stream margins, coastal areas.
   SOCIAL USES: *'Religious' uses*: inflorescences are rubbed on the arm with a piece of coconut cloth to improve fishing, Caroline Is. (Altschul 1973).

Scleria pterota *C. Presl* in Oken, Isis 21: 268 (1826).

Tropical Africa, Madagascar, S America; damp woodlands.

MEDICINES: *Genito-urinary system disorders*: rhizome-decoction is drunk for dysmenorrhoea, Tanzania (Burkill 1985); *Infections/infestations*: plant ash is given in small doses to infants for colds, Tanzania (Burkill 1985); 2 – 3 plants are cut up fine and cooked, the resultant mixture is given to cattle for rinderpest, Tanzania (Burkill 1985).

Scleria purdiei *C. B. Clarke*, Kew Bull. Addit. Ser. 8: 57 (1908).
Tropical S America; savannahs.

MEDICINES: *Unspecified medicinal disorders*: used in popular medicine, Venezuela (Pittier 1971).

Scleria secans (*L.*) *Urb.*, Symb. Antill. 2: 169 (1900).
Schoenus secans L., Syst. Nat. ed. 10: 865 (1759).
Central America, W Indies, tropical S America; forests and savannahs.

MATERIALS: *Fibres*: used to make fine paper, Brazil (Pio Corrêa 1926).

Scleria sylvestris *Poepp. & Kunth* in Kunth, Enum. Pl. 2: 346 (1837).
Tropical America; forest, savannnah.

MATERIALS: *Fibres*: used for making fine paper, Brazil (Pio Corrêa 1926); used for weaving chair seats, mats etc., Brazil (Pio Corrêa 1926).

Scleria sumatrensis *Retz.*, Observ. Bot. 5: 19, t. 2 (1789).
Dry open places and forests, also swampy areas; India and Sri Lanka to N Australia and Caroline Is.

MEDICINES: *Infections/infestations*: rhizomes made into a decoction for gonorrhoea, with *Pandanus* Parkinson spp. (*Pandanaceae*), *Areca catechu* L. (*Arecaceae*) roots and red chillies, Malay Peninsula (Burkill 1935).

WEEDS: *Perennial crops* (Kühn 1982); *Aquatic biotopes*: (Kühn 1982); *Forests*: (Kühn 1982).

Scleria tesselata *Willd.*, Sp. Pl. 4: 315 (1805).

var. **sphaerocarpa** *E. A. Rob.*, Kew. Bull. 18: 526 (1966).
Tropical Africa, Madagascar, India; wet, grassy places.

WEEDS: *Grassland*: (Kühn 1982); *Aquatic biotopes*: (Kühn 1982); *Rice fields*: (Kühn 1982).

TRICHOPHORUM *Pers.*

Trichophorum caespitosum (*L.*) *Hartm.*, Handb. Skand. Fl. 5: 259 (1849).
Scirpus cespitosus L., Sp. Pl. 1: 48 (1753).
C and N Europe, Siberia, N America; open, wet places.

ANIMAL FOOD: *Unspecified part*: grazed by sheep (Pratt 1900).

Trichophorum hudsonianum (*Michx.*) *Pers.*, Syn. Pl. 1: 70 (1805).
Eriophorum hudsonianum Michx., Fl. Bor.-Amer. 1: 34 (1803).

Scirpus hudsonianus Fernald, Rhodora 8: 161 (1906).
Europe, N central Asia, N America; rock crevices, open, wet peaty areas.
ENVIRONMENTAL USES: *Ornamentals*: (Huxley 1992).

UNCINIA *Pers.*

Uncinia divaricata *Boott* in Hook. f., Fl. Nov.-Zel. 1: 286 (1853).
New Zealand; open forest, scrub, river-bed, tussock grassland.
ENVIRONMENTAL USES: *Ornamentals*: (Grounds 1989).

Uncinia egmontiana *Hamlin*, Domin. Mus. Bull., Wellington 19: 33 (1959).
New Zealand; tussock grassland, scrub, bogs.
ENVIRONMENTAL USES: *Ornamentals*: (Grounds 1989; Huxley 1992).

Uncinia rubra *Boott* in Hook. f., Fl. Nov.- Zel. 1: 287, t. 64 (1853).
New Zealand; grassland, open scrub, bogs.
ENVIRONMENTAL USES: *Ornamentals*: (Grounds 1989; Huxley 1992).

Uncinia uncinata (*L. f.*) *Kük.* in Engl., Pflanzenr. 4 (20), 38 Heft: 62 (1909).
Carex uncinata L. f., Suppl.: 413 (1782).
New Zealand, U.S.A. (Hawaii); forest, scrub, bogs.
ENVIRONMENTAL USES: *Ornamentals*: (Grounds 1989).

TAXA ARRANGED ACCORDING TO FIRST AND SECOND LEVEL DESCRIPTORS

FOOD
Unspecified part: **Cyperus** articulatus, congestus, papyrus; **Schoenoplectus** americanus, lacustris.
Entire plant: **Scleria** biflora.
Aerial parts: **Cyperus** commixtus; **Gahnia** grandis.
Tubers: **Actinoscirpus** grossus; **Cyperus** bulbosus, congestus, cyperoides, esculentus, fulgens, usitatus; **Eleocharis** dulcis; **Fuirena** umbellata; **Lepironia** articulata.
Rhizomes: **Bolboschoenus** fluviatilis, maritimus; **Courtoisina** assimilis; **Cyperus** articulatus, compressus, distans, esculentus, gioli, hemisphaericus, maculatus, papyrus, rotundus, subumbellatus, victoriensis; **Eleocharis** sphacelata; **Pycreus** macrostachyos, muricatus; **Schoenoplectus** lacustris subsp. validus, littoralis.
Culms: **Courtoisina** assimilis; **Cyperus** amauropus, aucheri, maculatus, papyrus, rotundus, rotundus subsp. tuberosus, subumbellatus; **Kyllinga** erecta; **Schoenoplectus** corymbosus var. brachyceras, littoralis.
Leaves: **Courtoisina** assimilis; **Lepidosperma** gladiatum; **Scleria** pergracilis.
Nutlets: **Bolboschoenus** paludosus; **Carex** filicina; **Schoenoplectus** dissachanthus; **Scirpodendron** ghaeri.

FOOD ADDITIVES
Unspecified part: **Cyperus** alternifolius, distans, esculentus, exaltatus, haspan, laevigatus; **Fuirena** umbellata.
Entire plant: **Pycreus** nitidus.

Rhizomes: **Cyperus** alternifolius, subumbellatus; **Kyllinga** erecta, pumila.
Culms: **Kyllinga** squamulata.
Exudates: **Kyllinga** erecta, pumila, squamulata.

ANIMAL FOOD

Unspecified part: **Ascolepis** capensis; **Bolboschoenus** caldwellii; **Bulbostylis** barbata, hispidula; **Carex** albonigra, amplifolia, aquatilis, athrostachya, bella, bigelowii, brevior, brevipes, canescens, chlorosaccus, coriacea, demissa, divulsa, douglasii, ebenea, egglestonii, eleocharis, elynoides, eurycarpa, festivella, filifolia, foenea, geyeri, haydeniana, helophila, hepburnei, heteroneura var. epapillosa, heteroneura var. chalciolepis, hoodii, idahoa, illota, jonesii, kelloggii, lanuginosa, leporinella, luzulina var. ablata, lyngbyei, lyngbyei subsp. cryptocarpa, media, mertensii, microptera, miserablilis, multicostata, nebraskensis, nelsonii, nigra, nigricans, nova, obtusata, oreocharis, pachystachya, pachystylis, pelocarpa, petasata, phaerocephala, pityophila, praegracilis, praticola, preslii, pseudoscirpoides, pyrenaica, rariflora, raynoldsii, rossii, rostrata, rupestris var. drummondii, saxatilis var. major, scopulorum, scopulorum var. bracteosa, simulata, spectabilis, sprengelii, subfusca, subnigricans, sychnocephala, teneraeformis, tolmiei, vallicola, vernacula, vesicaria, exerantica; **Cyperus** albosanguineus, alterniflorus, amabilis, amauropus, arenarius, articulatus, aucheri, bulbosus, capitatus, castaneus, chordorrhizus, compressus, conglomeratus, conglomeratus var. effusus, dichroöstachyus, difformis, distans, dives, elatus, eragrostis, esculentus, exaltatus, fischerianus, fuscus, haspan, imbricatus, iria, jeminicus, laevigatus, ligularis, longibracteatus, longus var. pallidus, luzulae, maculatus, malaccensis, maranguensis, michelianus subsp. pygmaeus, obtusiflorus, papyrus, pilosus, pseudosomaliensis, rigidifolius, rotundus, rotundus subsp. merkeri, rotundus subsp. tuberosus, rubicundus, sphacelatus, squarrosus, strigosus, subumbellatus, tomaiophyllus, vaginatus subsp. gymnocaulos, zollingeri; **Desmoschoenus** spiralis; **Eleocharis** capillacea, flavescens, geniculata, mutata; **Fimbristylis** dichotoma, miliacea, ovata; **Fuirena** ciliaris, umbellata; **Gahnia** aspera, tristis; **Kobresia** capillifolia, pygmaea, schoenoides; **Kyllinga** brevifolia, erecta, nemoralis, odorata, pumila, squamulata, welwitschii; **Lipocarpha** chinensis, kernii; **Mapania** kurzii; **Nemum** spadiceum; **Pycreus** elegantulus, polystachyos; **Rhynchospora** barbata, cephalotes, hirsuta, holoschoenoides, nervosa subsp. ciliata, tenerrima; **Schoenoplectus** californicus, juncoides, lateriflorus, muricinux; **Scleria** foliosa, lithosperma, mitis; **Trichophorum** caespitosum.
Tubers: **Cyperus** esculentus, usitatus.
Rhizomes: **Cyperus** rotundus, usitatus.
Leaves/culms/aerial parts: **Carex** appressa, inversa, physodes; **Cyperus** clarus, esculentus, laevigatus, rotundus; **Eleocharis** dulcis, pallens, sphacelata; **Fimbristylis** dichotoma; **Remirea** maritima; **Schoenoplectus** corymbosus var. brachyceras, littoralis.
Inflorescences/infructescences/nutlets: **Cyperus** papyrus; **Fimbristylis** dichotoma.

BEE PLANTS: *Inflorescence*: **Cyperus** esculentus.

MATERIALS
Unspecified material type: **Cyperus** articulatus, corymbosus, corymbosus var.

longispiculatus; **Fimbristylis** dichotoma; **Kobresia** pygmaea; **Pycreus** mundtii; **Schoenoplectus** littoralis.

Fibres: **Actinoscirpus** grossus, grossus var. kysoor; **Afrotrilepis** pilosa, **Bolboschoenus** maritimus; **Bulbostylis** hispidula; **Carex** appressa, brizoides, dispalata, rhynchophysa, paniculata; **Cladium** mariscus, mariscus subsp. jamaicensis; **Cyperus** alopecuroides, alternifolius, articulatus, aucheri, canus, compactus, corymbosus, dichroöstachyus, digitatus, dives, elatus, esculentus, exaltatus, giganteus, grandis, haspan, imbricatus, immensus, iria, javanicus, laevigatus, latifolius, longus, longibracteatus, malaccensis, maranguensis, marginatus, natalensis, nutans, nutans subsp. eleusinoides, odoratus, pangorei, papyrus, procerus, procerus var. lasiorrhachis, prolifer, prolixus, sexangularis, subumbellatus, textilis, tomaiophyllus; **Desmoschoenus** spiralis; **Eleocharis** acutangula, calva, dulcis, elegans, interstincta, ochrostachys, sphacelata, spiralis; **Eriophorum** angustifolium, comosum; **Fimbristylis** dichotoma, ferruginea, spadicea, umbellaris; **Gahnia** trifida; **Hypolytrum** purpurascens; **Kyllinga** erecta, peruviana; **Lepidosperma** gladiatum; **Lepironia** articulata; **Machaerina** angustifolia, gunnii, mariscoides, rubiginosa; **Mapania** linderi, sumatrana; **Pseudoschoenus** inanis; **Pycreus** nitidus; **Rhynchospora** corymbosa, nervosa subsp. ciliata; **Schoenoplectus** americanus, californicus, corymbosus, corymbosus var. brachyceras, lacustris, lacustris subsp. validus, lateriflorus, littoralis, mucronatus, scirpoideus; **Schoenus** nigricans; **Scirpodendron** ghaeri; **Scirpoides** dioecus, holoschoenus, holoschoenus subsp. australis; **Scleria** boivinii, depressa, naumanniana, poaeformis, secans, sylvestris.

Cane/reed: **Eleocharis** geniculata, mutata; **Ficina** nodosa; **Lepidosperma** squamatum; **Schoenoplectus** riparius, tatora; **Schoenus** melanostachys.

Other materials/chemicals: **Bulbostylis** metralis, pilosa; **Carex** brizoides, paniculata, riparia; **Cyperus** articulatus, compressus, corymbosus var. scariosus, digitatus digitatus subsp. auricomus, distans, dubius, elegans, esculentus, glomeratus, haspan, iria, javanicus, ligularis, longibracteatus var. rubrotinctus, longus, longus var. pallidus, malaccensis, multifolius, rotundus, soyauxii, squarrosus, stoloniferus, subumbellatus, usitatus; **Eriophorum** angustifolium; **Kyllinga** erecta, pumila, squamulata; **Lagenocarpus** adamantinus; **Lepironia** articulata; **Pycreus** muricatus, nitidus; **Remirea** maritima; **Scleria** boivinii, depressa, naumanniana, poaeformis.

FUELS

Miscellaneous fuels: **Cyperus** papyrus.

Tinder: **Eriophorum** vaginatum.

SOCIAL USES

Smoking materials and drugs: **Cyperus** esculentus, prolixus.

Antifertility agents: **Cyperus** prolixus.

'Religious uses': **Afrotrilepis** pilosa; **Cyperus** alternifolius, articulatus, iria, latifolius, papyrus; **Hypolytrum** heteromorphum; poecilolepis; **Lipocarpha** chinensis; **Oxycaryum** cubensis; **Pycreus** mundtii; **Rhynchospora** corymbosa; **Scleria** boivinii, polycarpa.

Miscellaneous social uses: **Cyperus** articulatus, camphoratus, difformis, esculentus, imbricatus, laxus, longus, luzulae, odoratus.

VERTEBRATE POISONS

Mammals: **Carex** cernua; **Cyperus** cyperoides, longus; **Pycreus** nitidus; **Schoenoplectus** corymbosus.

NON-VERTEBRATE POISONS

Mollusca: **Cyperus** rotundus; **Schoenoplectus** sengalensis.

MEDICINES

Unspecified medicinal disorders: **Actinoscirpus** grossus, grossus var. kysoor; **Bolboschoenus** maritimus; **Bulbostylis** junciformis, lanata; **Carex** alboviridis, arenaria, disticha, hirta, macrocephala; **Cyperus** aggregatus, articulatus, corymbosus var. scariosus, elegans, esculentus, fastigiatus, haspan, laxus, longus, odoratus, prolifer, prolixus, rotundus, sexangularis, tenuiculmis; **Eleocharis** dulcis, geniculata; **Fimbristylis** miliacea; **Fuirena** umbellata; **Hypolytrum** compactum, laxum; **Isolepis** cernua; **Kyllinga** nemoralis, odorata, pumila; **Mapania** bancana; **Remirea** maritima; **Rhynchospora** nervosa subsp. ciliata, rubra; **Schoenoplectus** juncoides, lacustris, paludicola; **Scleria** boivinii, depressa, hirtella, pergracilis, purdiei.

Blood system disorders: **Bolboschoenus** yagara; **Carex** siderosticta; **Fimbristylis** umbellaris; **Kyllinga** brevifolia.

Circulatory system disorders: **Cyperus** articulatus, corymbosus var. scariosus, renschii, rotundus; **Scleria** iostephana.

Digestive system disorders: **Actinoscirpus** grossus, grossus var. kysoor; **Cyperus** alternifolius, articulatus, camphoratus, conglomeratus, corymbosus var. scariosus, esculentus, fuscus, longus, papyrus, prolixus, renschii var. scabridus, rotundus, rubicundus; **Kyllinga** brevifolia, nemoralis; **Rhynchospora** corymbosa; **Schoenoplectus** articulatus.

Endocrine system disorders: **Cyperus** rotundus; **Kyllinga** triceps.

Genito-urinary system disorders: **Actinoscirpus** grossus; **Bolboschoenus** maritimus; **Bulbostylis** capillaris, puberula; **Cyperus** articulatus, corymbosus var. scariosus, esculentus, longus, malaccensis, papyrus, rotundus, squarrosus; **Remirea** maritima; **Schoenoplectus** lacustris, wallichii; **Scleria** boivinii, depressa, foliosa, lithosperma, melanomphala, pterota.

Ill-defined symptoms: **Actinoscirpus** grossus; **Cyperus** articulatus, platyphyllus, serotinus var. inundatus; **Kyllinga** triceps; **Pycreus** mundtii.

Infections/infestations: **Actinoscirpus** grossus; **Cyperus** alternifolius, articulatus, bifax, camphoratus, compressus, conglomeratus, corymbosus var. scariosus, dilatatus, distans, esculentus, exaltatus, iria, javanicus, laxus, longus, rotundus, rotundus subsp. retzii, subumbellatus; **Eleocharis** dulcis; **Fimbristylis** junciformis, miliacea; **Kyllinga** brevifolia, erecta, nemoralis, triceps; **Mapania** cuspidata; **Rhynchospora** nervosa; **Scleria** boivinii, depressa, foliosa, pergracilis, pterota, sumatrensis.

Inflammation: **Cyperus** articulatus, corymbosus var. scariosus, maculatus, rotundus; **Fimbristylis** aestivalis; **Kyllinga** brevifolia, nemoralis; **Scleria** poaeformis.

Injuries: **Carex** nivalis; **Cyperus** alternifolius, cyperoides, distans, papyrus, subumbellatus; **Kyllinga** triceps; **Scleria** depressa.

Metabolic system disorders: **Cyperus** camphoratus, corymbosus var. scariosus, esculentus, longus.

Mental disorders: **Cyperus** articulatus, rotundus.

Muscular-skeletal system disorders: **Cyperus** articulatus, longus; **Fimbristylis** ovata.

Nervous system disorders: **Cyperus** corymbosus var. scariosus, iria, rotundus; **Kyllinga** triceps.

Nutritional disorders: **Actinoscirpus** grossus; **Cyperus** prolixus, rotundus.

Pain: **Bulbostylis** capillaris; **Cyperus** articulatus, corymbosus var. scariosus, cyperoides, distans, esculentus, rotundus; **Kyllinga** brevifolia; **Lipocarpha** chinensis; **Scleria** boivinii, depressa, naumanniana.

Poisonings: **Cyperus** articulatus, corymbosus var. scariosus, rotundus; **Kyllinga** brevifolia, nemoralis, triceps; **Schoenoplectus** junceus; **Scleria** boivinii.

Pregnancy/birth/puerperium disorders: **Cyperus** camphoratus, esculentus, exaltatus, longus, prolixus, rotundus; **Fimbristylis** dura, pauciflora; **Kyllinga** brevifolia; **Scleria** boivinii, lithosperma.

Respiratory system disorders: **Cyperus** articulatus, digitatus, rigidifolius, rotundus; **Fimbristylis** squarrosa; **Hypolytrum** purpurascens; **Pycreus** nitidus; **Schoenoplectus** juncoides; **Scleria** levis, lithosperma, pergracilis.

Sensory system: **Actinoscirpus** grossus; **Cyperus** esculentus, papyrus, rotundus; **Schoenoplectus** juncoides; **Scleria** boivinii.

Skin/subcutaneous cellular tissue disorders: **Actinoscirpus** grossus; **Bulbostylis** hispidula; **Cyperus** articulatus, cyperoides, luzulae, rotundus; **Fimbristylis** dichotoma; **Kyllinga** brevifolia, nemoralis, triceps.

ENVIRONMENTAL USES

Unspecified environmental uses: **Cyperus** conglomeratus.

Erosion control: **Carex** appressa, geyeri, hoodii, tolmiei, uruguiensis; **Cyperus** arenarius, articulatus, capitatus, chordorrhizus, conglomeratus, crassipes, jeminicus, rotundus, stoloniferus; **Desmoschoenus** spiralis; **Fimbristylis** ferruginea; **Fuirena** umbellata; **Remirea** maritima; **Schoenoplectus** californicus.

Revegetators: **Carex** exserta; **Caustis** dioica; **Cyperus** alopecuroides, chordorrhizus; **Desmoschoenus** spiralis; **Fimbristylis** cymosa; **Rhynchospora** corymbosa.

Soil improvers: **Actinoscirpus** grossus; **Fimbristylis** dichotoma, miliacea, pauciflora, schoenoides, umbellaris; **Fuirena** umbellata; **Rhynchospora** corymbosa; **Schoenoplectus** lateriflorus.

Ornamentals: **Bolboschoenus** maritimus; **Carex** acuta, acutiformis, alba, albula, appressa var. secta, arenaria, atrata, baccans, baldensis, berggrenii, brunnea, buchananii, caryophyllea, comans, conica, curvula, digitata, dipsacea, dissita, divulsa subsp. leersii, elata, firma, flacca, gaudichiana, grayi, hachijoensis, humilis, intumescens, kaloides, montana, morrowii, muskingumensis, nigra, ornithopoda, oshimensis, paniculata, pendula, petriei, phyllocephala, pilulifera, plantaginea, pseudocyperus, riparia, scaposa, siderosticta, sylvatica, testacea, trifida, umbrosa, uncifolia, vulpina; **Caustis** dioica; **Cladium** mariscus; **Cymophyllus** fraseri; **Cyperus** albostriatus, alopecuroides, alterniflorus, alternifolius, bulbosus, compressus, congestus, cyperoides, eragrostis, erythrorrhizos, esculentus, fertilis, haspan, longus, margaritaceus, owanii, papyrus, prolifer, sexangularis, ustulatus; **Desmoschoenus** spiralis; **Eleocharis** acicularis, dulcis, palustris, parvula, pusilla, vivipara; **Eriophorum** angustifolium, chamissonis var. albidum, latifolium, scheuchzeri, vaginatum, viridi-carinatum; **Gahnia** procera; **Isolepis** cernua, prolifera, setacea; **Machaerina** sinclairii; **Mapania** cuspidata, mannii, palustris; **Oreobolus** pectinatus; **Pycreus** filicinus; **Rhynchospora** colorata, nervosa, nervosa subsp. ciliata; **Schoenoplectus** tabernaemontani; **Schoenus** pauciflorus; **Scirpoides** holoschoenus; **Scirpus** cyperinus, sylvaticus; **Trichophorum** hudsonianum; **Uncinia** divaricata, egmontiana, rubra, uncinata.

WEEDS

Unspecified weed: **Cyperus** arenarius, articulatus, corymbosus var. scariosus, cyperinus, cyperoides, digitatus subsp. auricomus, dilatatus, distans, entrerianus, esculentus, haspan, grandibulbosus, iria, luzulae, papyrus, pilosus, pulcherrimus, rotundus, rotundus subsp. tuberosus, strigosus, subumbellatus, tenuiculmis, tenuis; **Fimbristylis** autumnalis; **Kyllinga** gracillima, nemoralis, polyphylla, squamulata; **Pycreus** flavidus; **Rhynchospora** nervosa subsp. ciliata; **Schoenoplectus** pungens, supinus.

Cultivation: **Bulbostylis** barbata, hispidula; **Cyperus** bulbosus, compactus, compressus, conglomeratus, cyperinus, cyperoides, difformis, digitatus, digitatus subsp. auricomus, distans, eragrostis, esculentus, iria, javanicus, malaccensis, mapanioides, maranguensis, michelianus subsp. pygmaeus, odoratus, rigidifolius, rotundus, rotundus subsp. retzii, rotundus subsp. tuberosus, soyauxii, sphacelatus, squarrosus, subumbellatus, tenellus, tenuiculmis, tenuis, tenuispica; **Fimbristylis** bisumbellata, miliacea; **Kyllinga** brevifolia, erecta, nemoralis, odorata, polyphylla, polyphylla var. elatior, pumila, squamulata.

Rotation crops: **Bulbostylis** barbata, densa; **Carex** albula, comans, panicea, **Cyperus** amabilis, articulatus, boreohemisphaericus, bulbosus, compressus, cuspidatus, cyperoides, dubius, esculentus, flavus, mapanioides, nutans, rigidifolius, rotundus; **Eleocharis** acutangula; **Fimbristylis** aestivalis, autumnalis, complanata, dichotoma, miliacea; **Fuirena** ciliaris; **Kyllinga** brevifolia, erecta, nemoralis; **Rhynchospora** nervosa subsp. ciliata; **Schoenoplectus** articulatus, juncoides; **Scirpoides** holoschoenus.

Perennial crops: **Bulbostylis** barbata, densa, hispidula; **Carex** brizoides, hirta, panicea; **Cyperus** articulatus, boreohemisphaericus, compactus, cyperoides, distans, esculentus, flavus, rotundus, squarrosus, zollingeri; **Fimbristylis** aestivalis, dichotoma, miliacea; **Kyllinga** brevifolia, nemoralis, triceps; **Rhynchospora** nervosa subsp. ciliata; **Scirpoides** holoschoenus; **Scleria** lithosperma, sumatrensis.

Grassland: **Bulbostylis** densa, hispidula; **Carex** albula, breviculmis, brizoides, buchananii, comans, coriacea, divulsa, hirta, inyx, longibrachiata, ovalis, panicea; **Cyperus** amabilis, compressus, cyperoides, difformis, dilatatus, distans, dubius, haspan, rotundus, sphacelatus, tenuiculmis; **Eleocharis** acicularis, geniculata; **Fimbristylis** dichotoma; **Fuirena** ciliaris, umbellata; **Kyllinga** brevifolia, erecta, nemoralis, triceps; **Lipocarpha** chinensis; **Pycreus** polystachyos, sanguinolentus; **Rhynchospora** nervosa subsp. ciliata; **Scleria** tesselata var. sphaerocarpa.

Waste places: **Bulbostylis** barbata, densa; **Carex** hirta, ovalis, paniculata, vulpinoidea; **Cyperus** albostriatus, alternifolius, amabilis, compactus, compressus, congestus, cuspidata, cyperoides, digitatus, distans, eragrostis, esculentus, flavus, iria, odoratus, radians, rotundus, squarrosus, subumbellatus; **Eriophorum** comosum;

Fimbristylis dichotoma, ferruginea; **Kyllinga** brevifolia, erecta, nemoralis; **Lipocarpha** chinensis; **Pycreus** polystachyos, sanguinolentus; **Rhynchospora** corymbosa; **Scirpoides** holoschoenus; **Scleria** lithosperma.

Disturbed land: **Cyperus** cyperoides, dilatatus, sphacelatus, tenellus; **Queenslandiella** hyalina.

Aquatic biotopes: **Bolboschoenus** caldwellii, maritimus; **Bulbostylis** densa, hispidula; **Carex** lasiocarpa, maorica, paniculata; **Cladium** mariscus; **Cyperus** alopecuroides, alternifolius, articulatus, compactus, congestus, cuspidata, difformis, digitatus, distans, haspan, imbricatus, iria, laevigatus, longus, malaccensis, odoratus, papyrus, pustulatus, radians, rotundus, serotinus, squarrosus, tenellus; **Eleocharis** acicularis, acutangula, atropurpurea, congesta, dulcis, elegans, geniculata, palustris; **Fimbristylis** acuminata, aestivalis, autumnalis, complanata, dichotoma, ferruginea, miliacea, umbellaris; **Fuirena** ciliaris, umbellata; **Lipocarpha** chinensis; **Oxycaryum** cubensis; **Pycreus** polystachyos, sanguinolentus; **Rhynchospora** nervosa subsp. ciliata; **Schoenoplectus** juncoides, lacustris, littoralis, mucronatus; triqueter; **Scleria** lithosperma, sumatrensis, tesselata var. sphaerocarpa.

Rice fields: **Actinoscirpus** grossus; **Bolboschoenus** maritimus; **Cladium** mariscus; **Cyperus** alternifolius, compactus, cuspidatus, cyperinus, cyperoides, denudatus, difformis, elatus, eragrostis, exaltatus, foliaceus, fuscus, haspan, imbricatus, iria, longus, longus var. pallidior, nutans, odoratus, pilosus, procerus, pustulatus, reduncus, remotispicatus, rotundus, serotinus, tenuispica; **Courtoisina** cyperoides; **Diplacrum** caricinum; **Eleocharis** acicularis, acutangula, congesta, dulcis, elegans, filiculmis, geniculata, mutata, philippinensis, retroflexa subsp. chaetaria; **Fimbristylis** acuminata, aestivalis, autumnalis, bisumbellata, complanata, dichotoma, miliacea, nutans, pauciflora, schoenoides, umbellaris; **Fuirena** ciliaris, stricta var. chlorocarpa, umbellata; **Kyllinga** polyphylla; **Lipocarpha** chinensis; **Pycreus** polystachyos, polystachyos var. laxiflorus, sanguinolentus; **Rhynchospora** corymbosa, holoschoenoides; **Schoenoplectus** juncoides, lateriflorus, mucronatus, roylei, supinus; **Scleria** lithosperma, tesselata var. sphaerocarpa.

Irrigation ditches: **Bolboschoenus** caldwellii; **Carex** maorica, ovalis, panicea; **Cyperus** eragrostis, exaltatus, longus var. tenuiflorus, rotundus, tenellus; **Schoenoplectus** roylei.

Gardens: **Carex** breviculmis, comans, inversa; **Cyperus** albostriatus, congestus, cyperinus, dilatatus, distans, eragrostis, esculentus, gracilis, laxus var. macrostachyus, mapanioides, obtusiflorus, michelianus subsp. pygmaeus, pseudosomaliensis, rigidifolius, rotundus, rotundus subsp. merkeri, sphacelatus, subumbellatus, tenellus, tenuiculmis; **Kyllinga** odorata, polyphylla; **Queenslandiella** hyalina.

Pasture: **Carex** comans, flagellifera, testacea, vulpinoidea; **Cyperus** eragrostis, latifolius, rigidifolius, ustulatus; **Kyllinga** polyphylla, triceps; **Pycreus** polystachyos, sanguinolentus.

Forests: **Carex** ovalis, sphacelatus; **Scleria** sumatrensis.

ACTIVE COMPOUNDS PRESENT

Alkaloids: **Bolboschoenus** maritimus; **Cladium** mariscus subsp. jamaicensis; **Cyperus** alternifolius.

ACKNOWLEDGEMENTS

We thank staff of the Centre for Economic Botany at Kew, particularly Hew Prendergast, Frances Cook, Steve Davis and James Morley for the help they have given to this project. We also thank Bob Allkin for his help and advice regarding the ALICE database system and Daniela Zappi for translating some of the references written in Portugese. Roger Milne assisted with data gathering from herbarium material. CAI acknowledges receipt of a Kew Sandwich Studentship during which the initial databasing work was carried out.

REFERENCES

Abbiw, D. K. (1990). Useful Plants of Ghana. Royal Botanic Gardens, Kew.

Adams, D. C., Staigmiller, R. B. & Knapp, B. W. (1989). Beef production from native and seeded northern Great Plain ranges. J. Range Manage. 42 (3): 243 – 247.

Adesina, S. K. (1990). Chemistry and biology of some Nigerian medicinal plants. Nigerian Field 55 (1 – 2): 71 – 80.

Adjanohoun, J. E., Ahiyi, M. R. A., Ake-Assi, L., Dramane, K., Elewude, J. A. *et al.* (1991). Contribution to ethnobotanical & floristic studies in western Nigeria. Traditional medicine and pharmacopoeia. Organization of African Unity, Scientific, Technical and Research Commission.

——, ——, ——, Alia, A. M., Amai, C. A. *et al.* (1993). Contribution to ethnobotanical & floristic Studies in Uganda. Traditional medicine and pharmacopoeia. Organization of African Unity, Scientific, Technical and Research Commission.

Ahmed, E. H. M., Bashir, A. K. & El Kheir, Y. M. (1984). Investigations of molluscicidal activity of certain Sudanese plants used in folk medicine. Part 4. Pl. Med. 1984: 74 – 77.

Altschul, S. S. P. von R. (1973). Drugs and foods from little-known plants; notes in Harvard University herbaria. Harvard University Press, Cambridge, Mass.

Al-Zoghet, M. (1989). Wild plants of Jubail & Yanbu. Royal Commission for Jubail and Yanbu, Jubail/Yanbu, Saudi Arabia.

Amalraj, V. A. (1991). Cultivated sedges of S. India for mat weaving industry. J. Econ. Taxon. Bot. 14 (3): 629 – 631.

Anon. (1988). Traditional Food Plants. Nutrition Paper 42. F.A.O., Rome.

—— (1990). Plants as Filters. Amer. Horticulturist 69 (3): 2.

—— (1994 – 2000). SEPASAL Database. Royal Botanic Gardens, Kew.

Arnold, T. H., Wells, M. J. & Wehmeyer, A. J. (1985). Khosian food plants — taxa with potential for future economic exploration. In: G. E. Wickens, J. R. Goodin, & D. V. Field (eds), Plants for arid lands: proceedings of the Kew International Conference on Economic Plants for Arid Lands held in the Jodrell Laboratory, Royal Botanic Gardens, Kew, England, 23 – 27 July 1984, pp. 69 – 86. Allen & Unwin, London.

Beaujard, P. (1988). Plantes et medecine traditionnelle dans le sud-est de Madagascar. J. Ethnopharmacol. 23 (2, 3): 165 – 265.

Bhattarai, N. K. (1993). Folk medicinal use of plants for respiratory complaints in central Nepal. Fitoterapia 64 (3): 243 – 259.

Burkill, H. M. (1985). The Useful Plants of West Tropical Africa 1 (2). Royal Botanic Gardens, Kew.

Burkill, I. H. (1935). A Dictionary of the economic products of the Malay Peninsula. Crown Agents for the Colonies, London.

Calanog, L. A. & Reyes, G. D. (1989). An integrated sustainable R & D program for CARP-ISF areas in the Philippines. Canopy Inter. 15 (4): 6 – 7, 11 – 12.

Campbell, A. (1986). The use of wild food plants, and drought in Botswana. J. Arid Environ. 11: 81 – 91.

Cauis, J. F. & Banby, J. (1935). The medicinal and poisonous sedges of India. J. Bombay Nat. Hist. Soc. 38 (1): 163 – 70.

Chin, Wee Yeow & Keng, Hsuan (1992). An Illustrated Dictionary of Chinese Medicinal Herbs. CRCS Publications, Sebastopol, California.

Chivinge, O. A. (1992). The effects of tillage methods on the survival of purple nutsedge (*Cyperus rotundus* L.) tubers. Discovery Innov. 4 (4): 75 – 77.

Cook, F. E. M. (1995). Economic botany data collection standard. Royal Botanic Gardens, Kew.

Dagar, H. S. & Dagar, J. C. (1986). Some observations of the ethnology of the Nicobareses with special reference to *Cocos nucifera* Linn. J. Bombay Nat. Hist. Soc. 83 (2): 306 – 310.

Dagar, J. C. & Dagar, H. S. (1999). Ethnobotany of Aborigines of Andaman-Nicobar Is. Suhrya International Publications, Dehra Dun.

Dangol, D. R. (1992). A review of weeds of maize fields in Nepal. J. Econ. Taxon. Bot. 16 (2): 489 – 500.

Deokule, S. S. & Magdum, D. K. (1992). Enumeration of medicinal plants from Baramati area, district Pune, Maharashtra state. J. Econ. Taxon. Bot. Addit. Ser. 10: 289 – 299.

De Vries, F. T. (1991). Chufa (*Cyperus esculentus, Cyperaceae*): a weedy cultivar or a cultivated weed? Econ. Bot. 45 (1): 27 – 37.

Dickoré, W. B. (1994). Flora Karakorumensis 1. Angiospermae, Monocotyledoneae. Stapfia 39: 1 – 298.

Dwivedi, S. N. & Pandey, A. (1992). Ethnobotanical studies on wild and indigenous species of Vindhyan plateau I. Herbaceous flora. J. Econ. Taxon. Bot. Addit. Ser. 10: 143 – 150.

Evans, D. K. (1991). Medicinal Cyperus (*Cyperaceae*) among the Shuar and Achuar of Southeastern Ecuador. Proc. West Virginia Acad. Sci. 63 (1): 8 – 9.

Fagotto, F. (1987). Sand dune fixation in Somalia. Environ. Conserv. 14 (2): 157 – 163.

Fosberg, F. R. (1988). Vegetation of Bikini Atoll. Atoll. Res. Bull. 315.

Fox, F. W. & Norwood Young, M. E. (1982). Food from the Veld. Edible Wild Plants of Southern Africa. Delta Books, Johannesburg & Cape Town.

Funk, E. (1978). Hawaiian fiber plants. Newsl. Hawaiian Bot. Soc. 17 (1/2): 27 – 35.

Goetghebeur, P. (1998). *Cyperaceae*. In: K. Kubitzki, H. Huber, P. J. Rudall, P. S. Stevens & T. Stützel (eds.), The families and genera of vascular plants 4: 141 – 190. Springer-Verlag, Berlin.

Gordon-Gray, K. D. (1995). *Cyperaceae* of Natal. Strelitzia 2. National Botanical Institute, Pretoria.

Greenway, P. J. (1950). Vegetable fibres and flosses in E. Africa. E. Afr. Agric. J. Vol. 15 (3): 146 – 153.

Grounds, R. (1989). Ornamental Grasses. Christopher Helm, London.

Haines, R. W. & Lye, K. A. (1983). The Sedges and Rushes of East Africa. East African Natural History Society, Nairobi.

Healy, A. J. & Edgar, E. (1980). Flora of New Zealand 3. P. D. Hasselberg, Wellington.

Heinsohn, D. (1990). Wetland plants as a craftwork resource. Veld & Flora 76 (3): 74 – 77.

Hermann, F. J. (1970). Manual of the Carices of the Rocky Mountains and Colorado Basin. U.S. Department of Agriculture, Forest Service. U.S. Department of Agriculture Handbook no. 374.

Heywood, V. H. (ed.) (1993). Flowering plants of the world. B. T. Batsford, London.

Hooker, W. J. (1857). Report on Vegetable Products Obtained Without Cultivation. London.

Huxley, A. (1992). The New Royal Horticultural Society Dictionary of Gardening. Royal Horticultural Society, Wisley.

Ingvason, P. A. (1969). The golden sedges of Iceland. World Crops 21: 218 – 220.

Irvine, F. R. (1957). Wild and emergency foods of Australia and Tasmanian Aborigines. Oceania 28 (2): ll3 – 142.

Jain, A. K. (1992). Ethnobotanical studies on 'Sahariya' tribals of Madhya Pradesh with special reference to medicinal plants. J. Econ. Taxon. Bot. Addit. Ser. 10: 227 – 232.

Jobe, R. T. (1991). The Niihau mat. Bull. Pacific Trop. Bot. Gard. 21 (3): 1 – 4.

Johnson, P. (1989). Wetland plants of New Zealand. DSIR Publishing, Wellington.

Jones, M. B. (1983). Papyrus — a new fuel for the third world. New Sci. 99: 418 – 421.

Kambu, K., Tona, L., Mulumba, M., Luki, N., Cimanga, K., Luzizila, Mukuna & Musambya, K. (1989). Etude chimique et biologique de tangawisi une boisson medicinale traditionnelle en Republique du Zaïre. Med. Trad. Pharm., Bull. Liason 3 (1): 3 – 13.

Kapur, S. K., Nanda, S. & Sarin, Y. K. (1992a). Ethnobotanical uses of RRL Herbarium — 1. J. Econ. Taxon. Bot. Addit. Ser. 10: 461 – 477.

——, —— & —— (1992b). Ethnobotanical uses of RRL Herbarium 2. J. Econ. Taxon. Bot. Addit. Ser. 10: 479 – 493.

Kern, J. H. (1974). Cyperaceae. In: C. G. G. J. van Steenis (ed.), Flora Malesiana ser. 1, 7 (3): 435 – 753.

Kühn, U. (1982). Monocot Weeds 3. Ciba-Geigy, Basle.

Kükenthal, G. (1935 – 1936). Cyperaceae — Scirpoideae — Cypereae. In: A. Engler (ed.), Das Pflanzenreich IV 20, 101 Heft. Engelmann, Leipzig.

Kukkonen, I. (1990). On the nomenclatural problems of Cyperus alternifolius. Ann. Bot. Fenn. 27: 59 – 66.

Kulhari, O. P. & Joshi, P. (1992). Fodder plants of Shekhawati region (Rajasthan). J. Econ. Taxon. Bot. Addit. Ser. 10: 355 – 370.

Leach, G. J. & Osborne, P. L. (1985). Freshwater plants of Papua New Guinea. University of Papua New Guinea Press, Port Moresby.

Lewington, A. (1990). Plants For People. Natural History Museum Publications, London.

Lord, T., Cubey, J., Grant, M. & Whiteley, A. (2000). RHS Plant Finder 2000 – 2001. Dorling Kindersley, London.

Manandhar, N. P. (1989). Medicinal plants used by Chepang tribes of Makawanpur district, Nepal. Fitoterapia 60 (1): 61 – 68.

Mathur, C. M. & Govil, D. P. (1987). Greening the desert. In: Desert Ecology; Scientific Reviews on Arid Zone Research 5. Scientific Publishers, Jodhpur, India.

Millar, A. G. & Morris, M. (1988). Plants of Dhofar, the southern region of Oman: traditional, economic and medicinal uses. Office of the Adviser for conservation of the Environment, Division of the Royal Court Sultanate of Oman.

Milliken, W. (1997). Plants for malaria, plants for fever. Royal Botanic Gardens, Kew.

Mohsin, A., Shah, A. H., Al-Yahya, M. A., Tariq, M., Tanira, M. O. M. & Ageel, A. M. (1989). Analgesic, antipyretic activity and phytochemical screening of some plants used in traditional Arab system of medicine. Fitoterapia 60 (2): 174 – 177.

Moore, L. E. & Edgar, E. (1970). Flora of New Zealand 2. A. R. Shearer, Wellington.

Mukhopadhyay, C. R. & Ghosh, R. B. (1992). Useful plants of Birbhum district, West Bengal. J. Econ. Taxon. Bot. Addit. Ser. 10: 83 – 95.

Muthuri, F. M. & Kinyamario, J. I. (1989). Nutritive value of Papyrus (*Cyperus papyrus, Cyperaceae*), a tropical emergent macrophyte. Econ. Bot. 43 (1): 23 – 30.

Navchoo, I. A. & Buth, G. M. (1992). Ethnobotany of Ladakh — J & K State. J. Econ. Taxon. Bot. Addit. Ser. 10: 251 – 258.

Nguyen, V. D. (1993). Medicinal Plants Of Vietnam, Cambodia & Laos. Privately published.

Noltie, H. J. (1993). Notes relating to the flora of Bhutan: XIX. *Kobresia* (*Cyperaceae*). Edinburgh J. Bot. 50(1): 39 – 50.

Pal, D. C. (1992a). Observation on folklore plants used in veterinary medicine in Bengal, Bihar and Orissa — 2. J. Econ. Taxon. Bot. Addit. Ser. 10: 137 – 141.

Pal, G. D. (1992b). Observations on less known interesting tribal uses of plants in lower Subansiri district, Arunachal Pradesh. J. Econ. Taxon. Bot. Addit. Ser. 10: 199 – 203.

Pandey, V. N. & Srivastava A. K. (1991). Yield and nutritional quality of leaf protein concentrate from *Eleocharis dulcis* (Burm. f.) Hensch. Aquatic Bot. 41 (4): 369 – 374.

Parsons, W. T. & Cuthbertson, E. G. (1992). Noxious Weeds of Australia. Inkata Press, Melbourne.

Partridge, T. R. (1992). Vegetation recovery following sand mining on coastal dunes at Kaitorete Spit, Canterbury, New Zealand. Biol. Conserv. 61 (1): 59 – 71.

Peters, C. R. (1992). Edible Wild Plants of Sub-Saharan Africa. Royal Botanic Gardens, Kew.

Pio Corrêa, M. (1926). Dictionário das plantas úteis do Brasil. Imprensa Nacional, Rio de Janeiro.

Pittier, H. (1971). Manual de las plantas usuales de Venezuela. Fundacion Eugenio mendoza, Caracas.

Plowman, T. C., Leuchtmann, A., Blaney, C. & Clay, K. (1990). Significance of the fungus *Balansia cyperi* infecting medicinal species of *Cyperus* (*Cyperaceae*) from Amazonia. Econ. Bot. 44 (4): 452 – 462.

Pratt, A. (1900). Flowering plants, grasses sedges and ferns of Great Britain 4. Warne, London.

Radanachaless, T. & Maxwell, J. F. (1994). Weeds of Soybean fields of Thailand. Multiple Cropping Center, Chiang Mai University.

Ratliff, R. D. & Westfall, S. E. (1992). Restoring plant cover on high-elevation gravel areas. California. Biol. Conserv. 60 (3): 189 – 195.

Reddy, M. B., Reddy, K. R. & Reddy, M. N. (1991). Ethnobotany of Cuddapah District, Andhra Pradesh, India. Int. J. Pharmocog. 29 (4): 273 – 280.

Rodin, R. J. (1985). The ethnobotany of the Kwanyuma ovambos. Missouri Botanical Garden, St Louis.

Rossetto, M., Dixon, K. W., Meney, K. A. & Bunn, E. (1992). In vitro propagation of Chinese puzzle (*Caustis dioica Cyperaceae*) — a commercial sedge species from Western Australia. Pl. Cell. Tissue Organ Cult. 30 (1): 65 – 67.

Saralamp, P., Chuakul, W., Temsirirkul, T. & Clayton, T. (1996). Medicinal plants in Thailand 1. Amarin Printing and Publishing Co., Bangkok.

Schermerhorn, J. W. & Quimby, M. W. (1960). The Lynn Index, Phytochemistry. Monograph 4, *Glumiflorae*. Massachusetts College of Pharmacy, Boston.

Shanmugasundaram, E. R. B., Mohammed-Akbar, G. K & Radha-Shanmuga-sundaram, K. (1991). Brahmighritham, an Ayurvedic herbals formula for the control of epilepsy. J. Ethnopharmacol. 33 (3): 269 – 276

Shukla, G., Mudgal, V. & Khanna, K. K. (1992). Notes on medicinal uses of plants known amongst rural folk-lore of Pratapgarh and Sultanpur districts, Uttar Pradesh. J. Econ. Taxon. Bot. Addit. Ser. 10: 219 – 225.

Simpson, D. A. (1992a). Notes on *Cyperaceae* in north-eastern Thailand. Cyperaceae Newslett. 10: 10 – 12.

—— (1992b). A revision of the genus *Mapania*. Royal Botanic Gardens, Kew.

—— (1993). Plant portrait: 225. *Rhynchospora nervosa*. Kew Mag. 10 (3): 117 – 121.

—— (1994). Plant portrait: 236. *Cyperus prolifer*. Kew Mag. 11 (1): 6 – 9.

—— (1996). *Cyperaceae*. In: M. J. E. Coode, J. Dransfield, L. L. Forman, D. W. Kirkup & I. M. Said (eds.), A checklist of the flowering plants and gymnosperms of Brunei Darussalam: 354 – 363. Ministry of Industry & Primary resources, Brunei Darussalam.

—— & Koyama, T. (1998). *Cyperaceae*. In: T. Santisuk & K. Larsen (eds.), Flora of Thailand 6 (4): 247 – 485. The Forest Herbarium, Royal Forest Department, Thailand.

Sudan Ministry of Co-op. and Rural Development (1980). Herbarium of Range Plants, Sudan.

Svenson, H. K. (1943). *Cyperaceae*. In: R. E. Woodson Jr. & R. W. Schery, Flora of Panama 2 (2). Ann. Missouri Bot. Gard. 30 (3): 185 – 229.

Täckholm, V. & Drar, M. (1950). Flora Of Egypt. 2. Bull. Fac. Sci. Fouad I Univ. 28: 1 – 475.

Tim, N., Tim, S. & Dunne, H. (1983). Getting to know Chinese vegetables. Pl. Gard. 39 (2): 4 – 19.

Tiwari, K. C., Joshi, G. C., Pande, N. K. & Pandey, G. (1992). Some rare folk tribal medicines from Garo hills in north eastern India. J. Econ. Taxon. Bot. Addit. Ser. 10: 319 – 322.

Toong, Y. C., Schooley, D. A., & Baker, F. C. (1988). Isolation of insect juvenile hormone III from a plant. Nature 333 (6169): 170 – 171.

Tournon, J., Raynal-Roques, A. & Zambettakis, C. (1986). Les Cyperacées médicinales et magiques de l'Ucayali. J. A. T. B. A. 33: 213 – 224.

Tredgold, M. H. (1986). Food Plants Of Zimbabwe. Mambo Press, Gweru, Zimbabwe.

USDA-ARS (2000). National Genetic Resources Program. Germplasm Resources Information Network — (GRIN) (Online Database). National Germplasm Resources Laboratory, Beltsville, Maryland.

USDA-NRCS (1999). The PLANTS database (http://plants.usda.gov/plants). National Plant Data Center, Baton Rouge, LA 70874-4490, U.S.A.

Van Eyk, B.-E. & Gericke, N. (2000). A guide to useful plants of southern Africa. Briza Publications, Pretoria.

Vedavathy, S. (1991). Antipyretic activity of six indigenous medicinal plants of Tirumala Hills, Andhra Pradesh, India. J. Ethnopharmacol. 33 (1, 2): 193 – 196.

Wickens, G. E., Haq, N. & Day, P. (eds.) (1989). New crops for food and industry. Chapman & Hall, London & New York.

LIST OF HERBARIUM SPECIMENS AND ARTEFACTS

All specimens are in K unless otherwise noted.

H1. *Randriamanantena & Durbin* 4282719 (MO)	H22. *El Hadidi* 12
	H23. *El Ghani* 4699
H2. *Randriamanantena & Durbin* 4282718 (MO)	H24. *Hosham & Mohammed* 29638
	H25. *Gillett* 9958
H3. *Randriamanantena & Durbin* 4282731 (MO)	H26. *El Ghani* 690 & 2204
H4. *Lousley* 3053	H27. *Collenette* 4299
H5. *Lambert & Thorp* 525	H28. *Keith* 1042
H6. *Brummitt* 6378	H29. *McClure* 13213 & 1389
H7. *Econmides* 1004	H30. *Shiu Ying Hu* 7174
H8. *Germeon* 1360	H31. *Maingay* 640
H9. *Mandaville* 296	H32. *Sampson* s.n.
H10. *Feinbrun & Zohary* 218	H33. *But* 79-127
H11. *Rechinger* 11703	H34. *Shiu Ying Hu* 5479
H12. *Haines* W297	H35. *Chung* 2805
H13. *Sahira* 38152	H36. *Sampson* 257
H14. *Hosham/Mohammed* 29639	H37. *Shiu Ying Hu* 7711
H15. *Zaraug* RMD34	H38. *Beil* 123A
H16. *Schwan* 9	H39. *Wight* s.n.
H17. *Miller & Long* 3317	H40. *Purseglove* P4060
H18. *Boulos* 17465	H41. *Purseglove* P4019
H19. *Trott* 388	H42. *Purseglove* P4057
H20. *Guest & Banaie* 25449 & 25450	H43. *Burkill* HMB2119
H21. *Watt* YEM B4	H44. *Simpson* 89/222
	H45. *Purseglove* P4059

H46. *Howard* 13
H47. *Kerr* 19673
H48. *Hennipman* 5025
H49. *Leano* 5426
H50. *Barber* 52
H51. *Ellen* 41
H52. *Merrill* 4247
H53. *Conklin & dol Rosario* 72631
H54. *Merrill* 162
H55. *Carr* 11427
H56. *Cruttwell* 100
H57. *Galore & Vandenberg* NGF41035
H58. *Coode, Cropley & Katik* NGF29747
H59. *Galore & Vandenberg* NGF41038
H60. *McKee* 1161
H61. *Blake* 1269
H62. *Johnson* 40153
H63. *Cross* 19165
H64. *Curtin* 40151
H65. *Blake* 18480
H66. *Koch* s.n.
H67. *Simon & Andrews* 2579
H68. *Hood* 40155
H69. *Jacobs* 674
H70. *Wilson* 5673
H71. *Hubbard* 2866
H72. *Koch* s.n.
H73. *Whistler* 6088
H74. *Powell* 49
H75. *Forbes* 481
H76. *Lyons* 27
H77. *Yuncker* 9939
H78. *Parham* 5607
H79. *Macrae* s.n.
H80. *Wit, Gbile & Daramola* 47376
H81. *Golding* 23
H82. *Adams* 3519
H83. *Terry* 3030
H84. *Terry* 3202
H85. *Gledhill* SL 1238
H86. *Deighton* 100
H87. *Morton* GC 6599
H88. *Haines* 43
H89. *Glenhill* 634
H90. *Wright* 48

H91. *Terry* 3045
H92. *Terry* 3030
H93. *Wickens* 3567
H94. *Bogdan* 5353
H95. *Haines* 4185
H96. *Haines* 4182
H97. *Guest, Rain & Rechinger* 16722
H98. *Guest* 25458
H99. *Thamer & Abdel Jaber* 50087
H100. *Goddard* s.n.
H101. *Howard* 12
H102. *Clarkson* 3842
H103. *Leach* BSIP19262
H104. *Schäfer* 5412
H105. *Yuncker* 9739
H106. *Christiansen* 54
H107. *Leach* BSIP19263
H108. *Macluskie* 3
H109. *Thornewill* 164
H110. *Deighton* 1417
H111. *Deighton* 929
H112. *Deighton* 5807
H113. *Gledhill* SL 2297
H114. *Dalziel* 825
H115. *Onochie* FHI32024
H116. *Barter* 1571
H117. *Dobson* 329
H118. *Ali Rawi* 16506
H119. *Boulos & Ads* 14044
H120. *Lambert & Thorp* 540
H121. *El Ghani* 2249
H122. *Guest* 1650
H123. *Davis & Coode* D39156
H124. *Simpson* 3143
H125. *Virgo* 46 & 49
H126. *Dickson* 489
H127. *Guest, al-Rawi & Abdul* 13535
H128. *Guest & al-Rawi* 14197
H129. *Dickson* 487
H130. *Dickson* 230
H131. *Wood* 72.07
H132. *Cope* 3
H133. *Suq ash-Shiyukh* 3305
H134. *Omar* 37756
H135. *Sahira* 38153
H136. *Daveau* s.n.

H137. *McClure* 7922
H138. *Hosie* 3
H139. *Koyama* 13987
H140. *Drummond* 46
H141. *Kerr* 18051
H142. *Santapau* 16722
H143. *Grierson & Long* 4384
H144. *Alston* 633
H145. *Kern* 8498
H146. *Mendoza* 97718
H147. *McKee* 1558
H148. *Whistler* W5326
H149. *Grierson & Long* 4385
H150. *Coode, Cropley & Katik* NGF29653A
H151. *Floyd* 3540
H152. *White* 13096
H153. *Blake* 11306
H154. *Koch* 358
H155. *Johnson* 26784
H156. *Koch* 391
H157. *Morton* A879A
H158. *Irvine* 1631
H159. *Adams* 4346
H160. *Morton* A1669
H161. *Keay* FHI37142
H162. *Haines* 44
H163. *Terry* 1803
H164. *Deighton* 5680
H165. *Terry* 3207
H166. *Leeuwenberg* 2126
H167. *Terry* 3003
H168. *Deighton* 4848
H169. *Deighton* 5676
H170. *Deighton* 1606
H171. *J. B. Hall* 2230
H172. *Wright* 46
H173. *Curry* 1083
H174. *Luke & Luke* 3733
H175. *de Lange* 215 (CHR)
H176. *Dalziel* 548
H177. *Deighton* 1417
H178. *Deighton* 1294
H179. *Deighton* 1295
H180. *Dalziel* 825
H181. *Onochie* FHI32024

H182. *Broadbent* 1
H183. *Fox* 123
H184. *Terry* 3030
H185. *Glover, Gwynne, Samuel & Tucker* 2316
H186. *Glover, Gwynne, Samuel & Tucker* 2561
H187. *Bogdan* 2616
H188. *Glover, Gwynne & Samuel* 185
H189. *W. J. J. O. de Wilde* 777
H190. *Terry* 3118
H191. *Adames* 123
H192. *Haines* 94
H193. *Ern, Hein & Pircher* 395
H194. *Morton* s.n.
H195. *Terry* 3255
H196. *Ern, Leuenberger, H. Scholz, U. Scholz & Schwarz* 1379
H197. *Tuley* 4
H198. *Hepper* 1243
H199. *Kershaw* 900426
H200. *Hall* 827
H201. *Vaillant* 2778
H202. *Cook* s.n.
H203. *Ern, Hein & Pircher* 701
H204. *Jordan* 316
H205. *Terry* 3132
H206. *Reekmans* 4038
H207. *Morton* GC8268
H208. *Geerling & Bokdam* 406
H209. *Morton* A1002
H210. *Morton* GC8004
H211. *Morton* A604
H212. *Morton* 6146
H213. *Barter* 620
H214. *Lowe* 2657
H215. *Broadbent* 110
H216. *Hepper* 3900
H217. *Tuley* 789
H218. *Callens* J73
H219. *Germain* 7749
H220. *Eden Foundation* 59
H221. *De Witte* 3196
H222. *Breteler et al.* 2358
H223. *Fay* 4508
H224. *Baudet* 452

H225. *Parker* E472
H226. *Terry* 3397
H227. *Andrews* A3586
H228. *Gillett & Hemming* 24489
H229. *Bally & Melville* 15264a
H230. *Harrison* H199
H231. *Harrison* 966
H232. *Myers* 6373
H233. *Andrews* A2703
H234. *Mooney* 6249
H235. *Parker* E175
H236. *Andrews* A749
H237. *McKinnon* S/261
H238. *Terry* 3348
H239. *Archbold* 62
H240. *National Irrigation Board* 20
H241. *Greenway* 2162
H242. *Glover, Gwynne & Samuel* 507
H243. *Okebiro* 702
H244. *Kaful* 63
H245. *Ruffo* 1442
H246. *Purseglove* P3736
H247. *Magogo & Glover* 576
H248. *Tanner* 1010
H249. *Hazel* 212
H250. *Glover, Gwynne & Samuel* 1130
H251. *Bogdan* 1901
H252. *Bogdan* 163
H253. *Tweedie* 517
H254. *Glover, Gwynne & Samuel* 434
H255. *Glover, Gwynne & Wateridge* 2946
H256. *MacDonald* 1289
H257. *Ombok* EA15858
H258. *Glover, Gwynne & Samuel* 453
H259. *Tanner* 3936
H260. *Hooper & Townsend* 2007
H261. *Glover, Gwynne & Samuel* 441
H262. *Faulkener* 964
H263. *Glover, Gwynne & Samuel* 470
H264. *Tanner* 4051
H265. *Michelmore* 1177
H266. *Mwangangi* 1430
H267. *Adamson* B3597
H268. *Glover, Gwynne & Samuel* 1203
H269. *Glover & Samuel* 3013
H270. *Glover, Gwynne & Samuel* 681

H271. *Glover, Gwynne & Samuel* 741
H272. *Glover, Gwynne, Samuel & Tucker* 1845
H273. *Smith* 1445
H274. *Mshasha* 131
H275. *Wild* 5073
H276. *Drummond* 5405
H277. *Angus* 2800
H278. *Fraser* 17
H279. *Burger* 1380
H280. *Hoare & Lupong* 49
H281. Economic Botany Museum Catalogue (EBMC) EMBC 34381
H282. EBMC 34393
H283. EBMC 34198
H284. EBMC 34193
H285. EBMC 73595
H286. EBMC 73609
H287. EBMC 40495
H288. EBMC 34392
H289. EBMC 34372
H290. EBMC 34370
H291. EBMC 34405
H292. EBMC 34404
H293. EBMC 34357
H294. EBMC 34396
H295. EBMC 34214
H296. EBMC 34315
H297. EBMC 34320
H298. EBMC 34239
H299. EBMC 34230
H300. EBMC 34235
H301. EBMC 34335
H302. EBMC 34352
H303. EBMC 34354
H304. EBMC 31740
H305. EBMC 34192
H306. EBMC 34195
H307. EBMC 34196
H308. EBMC 34200
H309. EBMC 34209
H310. EBMC 34229
H311. EBMC 34231
H312. EBMC 34236
H313. EBMC 34340
H314. EBMC 34346

H315. EBMC 34348
H316. EBMC 34355
H317. EBMC 34360
H318. EBMC 34361
H319. EBMC 34363
H320. EBMC 34364
H321. EBMC 34371

H322. EBMC 34372
H323. EBMC 34377
H324. EBMC 34383
H325. EBMC 34394
H326. EBMC 34429
H327. EBMC 73592
H328. EBMC 73793